Fieseler Fi 103R

David Myhra

Schiffer Military History
Atglen, PA

Acknowledgments

- *Mario Merino* - creator of the beautiful *Fieseler Fi 103R* digital images found in this illustrated history.

- *Ed Straten* - photographer who photographed a *Fieseler Fi 103R* in color at the Technical Museum, Delft, Holland. His dedication and persistence have given us these beautiful images of a faithfully restored *Fi 103R*.

Book Design by Ian Robertson.

Library of Congress Control Number: 2001091167

Printed in China.
ISBN: 0-7643-1398-3

We are interested in hearing from authors with book ideas on military topics.

Published by Schiffer Publishing Ltd.
4880 Lower Valley Road
Atglen, PA 19310
Phone: (610) 593-1777
FAX: (610) 593-2002
E-mail: Schifferbk@aol.com.
Visit our web site at: www.schifferbooks.com
Please write for a free catalog.
This book may be purchased from the publisher.
Please include $3.95 postage.
Try your bookstore first.

In Europe, Schiffer books are distributed by:
Bushwood Books
6 Marksbury Avenue
Kew Gardens
Surrey TW9 4JF
England
Phone: 44 (0) 20 8392-8585
FAX: 44 (0) 20 8392-9876
E-mail: Bushwd@aol.com.
Free postage in the UK. Europe: air mail at cost.
Try your bookstore first.

Fieseler Fi 103R

Several well-known *Nazi* German personalities claimed to be the inventor of the piloted *Fi 103 V-1*. The *Fi 103R*, as this machine came to be known, included at least two of *Adolf Hitler's* favorite people as its inventors...*SS Kommando Hauptsturmführer Otto Skonzeny*, the liberator of *Benito Mussolini* from his Italian prison cell high up in the Gran Sasso Mountains on September 12th, 1943, and *DFS* legendary test pilot aviatrix *Hanna Reitsch*. Their line of reasoning went like this: It was a common belief that Germany's war was lost unless a decisive blow could be struck against the Allies. *Skorzeny* and *Reitsch* believed that this could only be accomplished by the complete destruction of the eventual Allied assault upon the European Continent, such as what happened on D-Day, June 6th, 1944, at Juno, Omaha, Gold, and Sword beaches in Normandy, western France.

From this line of reasoning *Skorzeny* and *Reitsch* came up with the idea of a suicide corps. They thought that a weapon could be devised in the form of a flying bomb which, when piloted to its target, could sink a large warship, tanker, or troop transport. *Skorzeny* and *Reitsch* believed that enough of these piloted bombs could completely wreck any serious seaborne invasion with the expenditure of about 1,000 volunteer pilots. Both *Otto Skorzeny* and *Hanna Reitsch* were ready to volunteer. All they asked for was a weapon which would be certain to achieve its end, or to be allowed to convert one of the *Luftwaffe's* flying machines from its stable of military aircraft, such as the *Fieseler Fi 103 V-1* flying bomb.

The original *Fi 103 V-1* was a single pulse-jet engine powered unmanned missile. The term "*V-1*" came from the abbreviation of the word *Vergeltungswaffe Eins*, or "Revenge Weapon One." It has been suggested that the name came from the propaganda office headed by *Dr. Joseph Goebbels*, *Nazi* Minister of Propaganda, in order to bolster the sagging spirits of German population who had been promised secret weapons to turn the tide of war in their favor. Officially, to the German *Wehrmacht* the flying bomb was known as the *FZG 76* (*Flak Ziel Gerät*), that is, a long range target apparatus. The German Air Ministry (*RLM*), or *Reichsluftfahrtministerium* designation was the *Fieseler Fi 103*. It was the Americans who nicknamed it the "*buzz bomb*."

The father of the first practical pulse-jet engine like the one used to propel the flying bomb was the German fluid dynamicist *Dr.-Ing. Paul Schmidt*. The idea of the pulse jet was not a new one, even though he was granted a German patent (#523,655) for it on April 25th, 1931. Early pioneers into pulse-jet research included *Rene Lorin* (1908), *Marconnet* (1909), and *Rheinst* (1930), but it was *Paul Schmidt* who successfully applied the principles of the pulse-jet to propelling aircraft. Its first use in combat occurred on June 13th, 1944, on a flight across the English Channel. The British named this tiny machine "*doodlebug*," due to the highly distinctive sound of its single *Argus* As 014 pulse-jet engine.

Engineering work on the *V-1* was started in early 1942. The first recorded test flight of a *V-1* occurred on December 24th, 1942. All during 1943 little was done on the *V-1* while designers and constructors sought to prefect this very first cruise missile. In March 1944, the *V-1* was ordered into mass production. Essentially, the *V-1* was a small, single-wing airplane. It featured:

The manned *V-1* as envisioned by *SS Kommando Otto Skorzeny* and *DFS* test pilot *Hanna Reitsch*. Digital image by *Mario Merino*.

- A fuselage built up of six sections all fashioned out of welded 16-gauge mild steel sheet;
- Covering for the tail unit and controls were riveted to pressed steel ribs of about 22-gauge steel sheet;
- A nose formed the first fuselage section and was formed of an aluminum alloy cone of streamlined shape;
- At the tip of the nose cone was a small propeller by which the length of flight was governed. Air from the forward thrust caused it to rotate, and the number of rotations was noted by a counter inside which had been set before launching for a predetermined number, which when reached caused the controls to be reset from their flying position to dive onto its target;
- Immediately behind the course direction mechanism was a hollow wooden sphere or ball containing the compass used to guide the flight path.
- Monoplane wings fitted midway on the missile's fuselage;
- Its wings were squared off at their tips and measured 17 feet 6 inches in span;
- It had a tubular wing spar and metal ribs which were covered over with plywood sheeting;
- There were no ailerons in the wing;
- It had a conventional tail assembly consisting of a horizontal and vertical stabilizer with a rudder and elevators;
- The fuselage body of the *V-1* was cylindrical in shape (2 feet 7 inches at its greatest body diameter) and tapered sharply toward its stern;
- Nose to tail measured 21 feet 5 inches;
- The overall length of the missile (which included its extended tail section with the Argus As 014) was 25 feet 4 inches;
- Fully loaded the *V-1* weighed 4,750 pounds;

Four piloted models of the *Fi 103* were constructed for the *Reichenberg Program*, three of them being training versions:

Reichenberg 1 - a single-seat glider trainer version ...converted from the standard *FZG 76*, but without the *Argus As 014* pulse jet engine installed...equipped with a one piece cockpit canopy, water ballast in the nose area equal to the weight of the 1,874 pound *Amatol* warhead, landing skid, aileron, and flaps;

Reichenberg 2 - a dual cockpit trainer version of the *Reichenberg 1* converted from the standard *FZG 76*, but without the *Argus As 014* pulse jet engine installed. The second cockpit was located where the warhead would normally have been in the *FZG 76*, ailerons, no water ballast, but the tail fin and rudder area were enlarged over that of the *Reichenberg 1*;

Reichenberg 3 - a single-seat trainer version converted from the standard *FZG 76*, but equipped with a fully operational *Argus As 014* pulse-jet engine...cockpit canopy located directly in front of the *FZG's* air intake, water used as ballast to simulate the weight of 1,874 pound *Amatol* warhead, ailerons, landing skid, and flaps;

Reichenberg 4 - the operational version...converted from the standard *FZG 76*. It would have been taken to the vicinity of its target beneath a *Heinkel He 111H* in similar fashion to the pilotless *Fi 103s* launched against the British by *Kampfgeschwader 33*. Fitted with a foresight to assist in aiming towards the target.

***Otto Skorzeny*, it is remembered, was the man *Hilter* called upon to rescue Italian dictator *Benito Mussolini* from imprisonment in a hotel, 6,500 feet up in the *Gran Sasso* mountains in central Italy on September 12th, 1943. *Skorzeny* claimed to have had the idea first of converting a standard *V-1* flying bomb into a manned *V-1* so that the machine would have better assurance of hitting its target.**

Fi 103 Reichenburg 4
Dimensional Data
- Length, overall - 26 feet 3 inches
- Wing span - 18 feet 9 inches
- Length, fuselage - 24 feet 3 inches
- Diameter, fuselage 2 feet 9 inches
- Length, *Argus As 014* pulse jet propulsive unit - 12 feet
- Diameter, *Argus As 014* pulse jet propulsive unit - 1 foot 3 inches
- Length, each wing - 8 feet 1/3 inch
- Wing, width with aileron - 3 feet 6 inches
- Aileron, width - 9 3/4 inches
- Aileron, length - 7 feet 8 inches
- Horizontal stabilizer with control surface - 6 feet 9 inches by 2 feet 1 inch
- Horizontal stabilizer control surface (each) - 2 feet 10 inches by 9 inches
- Vertical stabilizer with control surface - 3 feet 2 inches by 1 foot 9 inches
- Vertical stabilizer control surface - 1 foot 4 inches by 1 foot nine inches
- Cockpit opening length - 1 foot 9 inches
- Cockpit opening width - 1 foot 5 inches
- Seat to top of canopy - 3 feet 1 inch
- Windshield length - 11 inches
- Windshield width at top - 6 inches
- Windshield width at bottom - 5 inches
- Length from nose to headrest - 14 feet
- Length of fuel cell - 4 feet 6 inches
- Length of war head - 6 feet
- Length from nose to wing - 7 feet 1 inch
- Length of wing spar - 14 feet 4 inches
- Diameter of wing spar - 4 3/8 inches

Technical Data

• The *Fi 103R 4s* were intended to be launched from bomber type aircraft. No information gathered to date suggests that the *Fi 103R 4s* were to be launched from ramps.

• The *Fi 103R 4* was to be flown and directed into the target by the pilot, and a parachute jump was to be made just before colliding with the target.

• No information has yet been discovered which indicates that the *Fi 103R 4* was intended for ramming.

• The *Fi 103R 4s* were constructed of steel except for the wings and nose cover, which were plywood. Structural parts of the *103R* are identical to the late models of the *V-1* flying bomb except for the following:

- ailerons added to standard wings;
- cockpit, controls, and flight instruments installed just forward of the *Argus As 014* ramjet unit;
- tapered war head with two fuse cells and plywood cover substituted for standard olive-shaped war head;
- only one compressed air container is used, and this was located at about the position of the automatic pilot in the standard *V-1*;
- instrument panel which included:
  - arming switch operated by key wired to instrument panel;
  - clock;
  - airspeed indicator;
  - altimeter;
  - combined inclinometer and turn indicator;
- a gyro compass was mounted in a shock mounted bracket with a small 24-volt wet battery and 3-phase inverter. This assembly was mounted on the floor between the pilot's knees, so that the gyro compass was just below the instrument panel;
- flight controls are the conventional stick and pivoted cross bar rudder control;
- personal equipment consisted of the following:
  - parachute;
  - life-preserver;
  - helmet with headphones;
  - throat microphones;
  - sunglasses;
  - safety belt;
  - shoulder straps;
  - crash-pad above instrument panel;
  - plywood bucket seat;
  - padded headrest at the top of seat back;

***Hanna Reitsch* claimed that she was the one who first thought of converting the *V-1* flying bomb into a pilot guided cruise missile for use against Allied shipping, such as troop carriers, tankers, and even battleships.**

Performance

Performance of the *Fi 103R 4* was expected to be the same as that of the late models of the standard *V-1* flying bomb. Performance specifications of the *V-1* include:

- takeoff speed leaving the takeoff ramp was 250 miles per hour [400 kilometers/hour];
- maximum speed of late model *V-1* was 555 miles per hour [900 kilometers/hour];
- time in flight about 40 minutes;

Pilot Survival With the *Fi 103R-4*

Although the pilots were to be equipped with parachutes, it was expected that 99% would not survive a bail out.

To bail out of the *Fi 103R-4* the pilot would have to remove the cockpit canopy, which was mounted near the intake of the *Argus As 014* pulse-jet engine. The canopy was mounted by two curved prongs which hinged into sockets on the right of the cockpit. The left side of the canopy was held by two eyelets and sliding pins. To release the canopy it was necessary to operate the lever on the left side of the cockpit, disengaging the sliding pins from their eyelets. The canopy then had to swing about 45 degrees before the front prong on the right side would release. The back edge of the canopy interfered with the cowling of the *Argus As 014* pulse jet unit. It was believed that it would be difficult, if not impossible, to jettison the canopy while in flight.

Lines on the windshield were drawn to estimate diving angles to the target.

Communication between launching aircraft and the pilot in the *Fi 103R-4* was maintained through a four channel connector at the top of the fuselage in front of the canopy.

Part of the Allied invasion fleet assembled at the south of France in August 1944. Both *Skorzeny* and *Reitsch* believed that the invasion could be stopped by manned *V-1s* being guided down into the sea borne troop carriers.

Quantity of *Fi 103R-4s* Produced and Shipped

Series production of the *Fi 103R-4s,* which were conversions of the original *Fi 103 V-1,* was to have been done by aircraft manufacturer *Gerhard Fieseler Werke* located in Kassel. A total of 175 brand new *Fi 103R-4s* were found at the Dannerberg assembly plant. Another one hundred twenty-five were found in storage buildings at the assembly plant's rail head ready to be loaded on to railroad cars. No *Fi 103R-4s* were found loaded on railway cars, and no information obtained to date indicates that any *Fi 103R-4s* had been shipped.

Characteristics of *Hanna Reitsch's* Personality

- Born March 29th, 1912, in the City of Hirschberg, province of Silesia, Germany;
- The daughter of wealthy parents...father a medical doctor specializing in eye surgery and had his own clinic;
- Attended private all-girl schools;
- Started flying (soaring) lessons in the Summer of 1930 at *Wolf Hirth's Grunau* Gliding School...she was their first female student;
- First female in Germany to pass the "C" level requirements for full certification as a glider pilot—Summer of 1931;
- Taught herself aero-engine engineering;
- Proof-read *Wolf Hirth's* book manuscript on the principles of thermal flying;
- Entered medical school in Berlin Fall of 1931;
- Took powered aircraft pilot training at the Staaken airfield outside of Berlin;
- Hired by Wolf Hi*rth* to be a summer workshop assistant at his *Grunau* Gliding School working in the repair of and constructing gliders;
- Entered the 1933 summer sailplane competitions at the *Rhön/Wasserskuppe* but failed to place;

- Invited by *Professor Walter Georgii*, leader of the International Study Commission for Motorless Flight, to join a privately financed scientific expedition to study thermal conditions in South America;
- Quit medical school (2nd year) in Fall of 1933 to become a full-fledged pilot;
- Raised money to join in the South American expedition by performing as a stunt pilot in a German movie titled "*Rivals of the Air*."
- Left for Rio de Janeiro, Brazil, in January 1934 and put on public sailplane flying demonstrations;
- Returned to Germany from Brazil in April 1934;
- Went to Finland in September 1934 to participate in sailplane flying demonstrations;
- Represented Germany at the Spring 1935 International Air Display, Lisbon, Portugal;
- *Reichs Air Ministry* invited her to attend airline pilot's school at Stettin, becoming the world's first female commercial pilot student and Germany's first female commercially licensed pilot;
- Invited by *Professor Walter Georgii* to join a scientific expedition to North Africa in the Summer of 1935 to study thermal wind currents;
- World War II begins September 1939;
- Invited by *Professor Walter Georgii* to join his *DFS* at Darmstadt-Griesheim as a test pilot in the Flight Department, flight testing gliders, training sailplanes, high-performance aircraft, and special purpose aircraft;
- Although on loan for long periods of time to the *Luftwaffe*, always considered a civilian employee of the DFS, one of the RLM's research institutes up until Germany's surrender in May 8th, 1945. She was never a member of the *Luftwaffe*, nor did she hold any military rank;
- Set a new world record for long-distance soaring of over 100 miles;
- Test flew the world's first high-performance amphibious glider, the "*Sea Eagle*," on Lake Bodensee in southern Germany;
- Test flew the *Madelung* catapult launching device designed to assist the takeoff of heavily loaded transports from the water;
- Test flew the *Hans Jacobs*-designed dive brakes on a *Sperber* glider in 1936 starting a dive at 19,000 feet and pulling out at 600 feet...the dive brakes held steady and became a standard feature on *Stuka* dive bombing aircraft;
- Nominated in 1937 by *General Ernst Udet* for the title of "*Flugkapitän* (flight captain), becoming the first female to be honored with this title, and was presented to *Hitler* for the first time;
- One of a group of five people to be first to fly over the Alps in a sailplane in Spring 1937;
- Asked by *Udet* in September 1937 to test dive brakes designed into new *Luftwaffe* fighter and bomber aircraft at the *Luftwaffe Testing Center*, Rechlin, however, fellow male test pilots resisted and *Generals Ernst Udet* and *Ritter von Greim* had to use their high rank to intervene;
- Test flew *Professor Focke's Focke-Achgelis Fa-61*, considered to be the world's first successful vertically rising aircraft;
- *Hermann Göring* awarded her the meritorious *Luftwaffe's* "Military Flying Medal," becoming Germany's first female to do so...*Göring* also gave one to American *Charles Lindbergh* at the same ceremony;
- February 1938 demonstrated the *Focke-Achgelis Fa-61* (first practical helicopter) inside the International Auto Show held at Berlin's *Deutschlandhalle* Stadium;

Normandy, France, D-Day, June 6th, 1944. The Allies are pouring ashore, and the *Luftwaffe* is nowhere to be seen.

- July/August 1938 participated in the International Air Races being held at Cleveland, Ohio, taking *Ernst Udet's* place...putting on a demonstration of aerobatic flying as part of the air show;
- *New York Times* of August 23rd, 1938, welcomes *Hanna Reitsch* to New York City "German Girl Flier Here. *Miss Reitsch*, Women's Glider Champion, Going To Air Races;"
- Test flew *DFS's* prototype *DFS 230* cargo- and troop-carrying glider, which was later used by *Wehrmacht* paratroopers to storm the Belgian *Fort Ebel Emael* and forts on the island of Crete;
- Got inside a prototype gasoline tank towed behind bomber aircraft to witness and experience characteristics in flight. In her ten years of aircraft testing this particular task produced her greatest strain...both airsickness and claustrophobia;
- Flew a small observation plane into a snare in order to determine if aircraft observation planes could land on cable snares strung on a warship;
- Test flew a *Dornier Do 17* rigged with a cable-cutting device into moored barrage balloons of the type found over England. The *Luftwaffe* was able to determine that cable cutters worked well. *Ernst Udet* recommended her for the *Iron Cross 2nd* class and she was presented with it by *Hitler* on March 28th, 1941. This is her 2nd meeting with *Hitler*, and she is the 1st German female to win the award in World War II;
- Test flew the *Messerschmitt Me 163B* bi-fuel liquid rocket fighter in October 1942. She was called upon to test fly the "*Komet*" after *Heini Dittmar* suffered injury in test flying the *Me 163.* On her 5th test flight the takeoff dolly failed to fully release and all electrical systems went dead. She crash-landed the *Me 163*, suffering severe head injuries, and remained in the hospital for five months. *Hitler* awarded her the *Iron Cross 1st* class during her convalescence;
- Resumed flight testing in the Summer of 1943 with the *Messerschmitt Me 328* and set world altitude speed records in *Me 163s*, but due to war-time secrecy, they were never made public;
- November 1943 flew *General Robert Ritter von* Greim, commander of *Luftwaffe Air Fleet #6* to the Russian Front, spending three weeks visiting *Luftwaffe* units assigned there and tasting what it was like to be shelled by enemy artillery;
- Suggested a manned/piloted *V1* for increased and improved aiming and guidance when aimed at Allied shipping and supply trains. She believed that pilots of manned *V-1s*, by sacrificing their own lives, would save many times that number of their fellow countrymen. Test flew a modified *V1* known as *Fi 103R*, but the concept did not have wide appeal...in fact, there were many high-ranking critics, such as *Milch, Göring, Hitler*, and others. When the kamikaze-type project appeared workable, delays in supplies and equipment saw its military targets, especially the Normandy seaborne invasion, vanish, thus its effective military moments were lost;
- Severely injured during a late 1944 Allied air raid on Berlin;
- Told by friends that Hitler was exterminating Jews... she believed that it was all Allied propaganda;
- April 25th, 1945, *Hitler* telephones. Berlin is encircled by the Red Army and they are shelling the city. He ordered *General Ritter von Greim* to his bunker beneath the *Reich Chancellery* to tell him that he was now chief of the *Luftwaffe*. He has fired *Göring*. She pilots an airplane 60 miles, landing in Berlin, and spent three days in the *Fürhrer* bunker. *Hilter* at first orders them to stay, then tells them to leave. *Hanna Reitsch* flies herself and *von Greim* out of Berlin in an *Arado Ar 96B* advanced trainer at night as Soviet ground forces try to shoot them down;
- Arrested postwar, her friend and mentor *General Robert Ritter von Greim* commits suicide late May 1945 in a POW camp in Germany;
- Arrested postwar by American Military authorities in October 1945, she is held in captivity for 15

British troops on shore at Juno, Gold, and Sword beaches D-Day, June 6th, 1944. At the end of the day they had advanced 5 miles inland.

months imprisoned at a POW camp at Oberursel, near Frankfurt am Main;
- Wrote her autobiography "*Flying Is My Life,*" and an English version appeared in 1954;
- Visited *Nehru* in India in the 1950s, teaching him and members of his family to fly;
- Spent several years in Ghana helping the government organize the Ghanian Air Force and training glider pilots;
- Member of the elite world-wide Experimental test Pilot's Association;
- Wrote her memoirs in 1975 and published in a book titled "*The Sky My Kingdom,*" viewing the *Third Reich* more critically than before;
- Died Frankfurt am Main August 24th, 1979, age 67;

What The Critics Said Of *Hanna Reitsch*:

- Big fan of *Hitler*;
- Became friendly with *Hitler*...she had four audiences with him;
- The last individual to speak personally with *Hitler* in his Berlin *Führer* bunker only hours before he and *Eva Braun* committed suicide, and then flew to Rechlin from Berlin to rally the *Luftwaffe* to continue fighting;
- An eyewitness to the fall of the *Third Reich*;
- True believer in *Nazism* to the very end;
- Within *Nazism* she found opportunities and rewards;
- Her two books written post war are defensive and self-serving;
- Doggedly unrepentant...to the very end of her life;
- Learned from friends that *Hitler* and *Himmler* were exterminating Jews...*Himmler* assured her that it was all Allied propaganda and she believed him;
- Sought desperately to rally the *Luftwaffe* and save the *Reich*;
- Generous in helping other woman pilots from other countries;
- Very proud and very brave individual;
- A woman who excelled in a government environment (*Nazism*) that was for the most part extremely repressive;
- Achieved personal success in a culture which defined women in terms of wife and mother, and lived her passion for flying;
- Consorted with the devil (*Hitler* and *Nazism*) to establish opportunities and rewards for her achievements;
- Excited about the proposed manned *Fi 103R-4 V-1 Kamikaze*-type bomb and its suicide pilots... disappointed when the project was abandoned;
- Excellent test pilot ranked among the world's best;
- Excellent glider pilot, with approximately 40 sailplane records to her credit;
- Accomplished her last world record at age 65, two years before she died;

The Piloted V-1...Otto Skonzeny in His Own Words

One day I had the opportunity to visit Peenemünde and witness the launching of one such *V-1*. I flew with an engineer colonel of the *Luftwaffe*, who was a specialist in these flying bombs, and on the return flight I discussed with him the question: would it not be possible to have the *V-1* flown by a pilot?

The very evening of that summer day in 1944 we set to work, together with *Focke-Wulf* and *RLM* engineers. I had invited them to a villa on the Wannsee. A dozen engineers began drawing plans...on the billiards table and even lying on the floor; we had to find sufficient room in the *V-1* to accommodate a pilot with ejection seat and parachute.

We worked the whole night, and by morning we had the solution. All we had to do was build a prototype. *Feldmarschall Milch* gave me the "clear road," provided that an *RLM* commission raised no questions. The chairman of this commission was a venerable admiral. "Where do you intend to get the workers, foremen, and engineers to build this prototype? We do not have enough labor forces as it is, especially in the aviation industry."

I replied that near Friedenthal there was a *Heinkel* factory that was not operating at full capacity. *Professor Ernst Heinkel* had personally offered me three engineers and five mechanics, and as well had placed three empty work barracks at my disposal.

"Good," said the admiral, "but you can only carry out your work with already built *V-1s*, and you know that we have none."

That is not what *Professor Porsche*, a friend of mine, said to me. There are several hundred *V-1s* in his *Volkswagen* factories waiting to be picked up. I can assure you that he would gladly let me have a dozen.

The intended use of the manned *V-1* as *Hanna Reitsch* and *Otto Skorzeny* envisioned would be similar to the Japanese "*kamikaze*" pilots. Seen is a "*kamikaze*" about to slam into the port edge of the U.S. Navy aircraft carrier *Essex (CV-9)*. This "*kamikaze*" attack on November 25th, 1944, killed 15 sailors and wounded 44.

In a short time I had two small workshops at *Heinkel-AG* Friedenthal. I had tables and beds moved in. Everyone, engineers, foremen, and workers, worked at full speed, sometimes more than 14 hours a day, in order to bring our so-called "*Operation Reichenberg*" to fruition as quickly as possible.

When I saw *Feldmarschall Milch* again, he smiled. "Well then, *Skorzeny*, satisfied hopefully?"

"Naturally," I answered him, "in spite of the 2 to 3 week delay."

"Three weeks in such a project, that's nothing. A manned *V-1*! If you can roll out your prototype in 4 to 5 months I'll congratulate you again."

"*Herr Feldmarschall*, I hope that I can show you the prototype in 4 to 5 weeks!"

He looked at me seriously and thought that I was making a joke. Then he shook his head.

"You are deluding yourself, my dear fellow. That is all well and good. But do not make too much of it. We will talk about this machine again in 4 to 5 months. Until then, lots of luck."

Our workshop at *Heinkel*-Friedenthal was actually a craftsman operation, but one that worked with success. When I could I spent several hours each day in "my factory." After 14 days I again contacted *Feldmarschall Milch* and informed him that we had both been wrong: I had three *V-1s* ready to test fly.

Feldmarschell Milch was amazed. He gave me authorization to undertake three takeoff attempts at Gatow airfield. Two test pilots were chosen. The manned *V-1* was not launched from ramp like the original *V-1* was; instead, the manned *V-1* was to be towed by a *Heinkel He 111H* to a height of 2,000 meters (6,562 feet) and then released. Both manned *V-1* machines made crash landings, however, both pilots escaped with injuries.

A downright dour *Feldmarschall Milch* told me that a commission would be appointed to investigate the causes of the bad landings.

For the time being I was forbidden to make any further attempts. I was speechless. Had we worked too carelessly and too quickly?

Then *Hanna Reitsch*, our legendary female test pilot, telephoned me. Since her serious crash in an *Me 163B*, October 30th, 1942, she told me that she had had the same idea as I several months earlier...the *V-1* could be flown as a manned aircraft! But she had received the official order to drop the idea. She said that there was no need to wait for the results of the investigation to learn the causes of our two accidents: both pilots had previously flown only propeller-driven aircraft. Our prototype, which was much lighter than a standard *V-1*, reached a speed of 700 kilometers per hour (435 miles per hour) and a landing speed of 180 kilometers per hour (112 miles per hour). All this speed made the pilots more than uncertain when it came time to land. *Hanna* and two of her associates who had likewise flown jet and rocket powered aircraft, declared themselves ready to repeat the attempt.

I declined firmly and reminded them of the official, strict order from *Milch* that he would not make an *He 111H* available to us at Gatow airfield. She shrugged her shoulders and said, "I took you for a man who is willing to take a chance! One can always fly if only one wants to. My friends and I have visited your workshop and examined your first manned *V-1s*. I am sure that we are not fooling ourselves: they are outstanding aircraft. We will talk more about it later. Until tomorrow."

I must admit that I could not close my eyes that night. A third accident would be unimaginable. Did I have the right to plunge this wonderful aviatrix into

***Hanna Reitsch* and *Otto Skorzeny* strongly believed that this was how the piloted *V-1* should be employed against Allied shipping.**

such an adventure? The next day *Hanna Reitsch* and her two companions were so convincing that I took it upon myself to dupe the airfield commander. I acted completely natural and told him that I had just received approval to continue "*Operation Reichenberg*." I asked him his opinion on several questions and assigned two of my officers not to let him out of their sight, to accompany him into the mess and to take care that he did not telephone *Feldmarschall Milch's* staff. When I saw the *V-1* flown by *Hanna Reitsch* separate from the *He 111H*, my heart pounded as never before. She had taken full responsibility onto herself without hesitation. She knew that her airspeed on landing would be about 180 kilometers (112 miles per hour). I was firmly convinced, however, that she would pull it off. And she did! She landed smoothly and then repeated the test flight. I congratulated her with all my heart. "That is a wonderful aircraft," she said to me, "we will be able to do something with it."

The other two test pilots also flew the manned *V-1*, and both landed without any difficulty. The manned *V-1* was not destined to be a success, however.

When the flights by *Hanna Reitsch* and her two companions became known, we received permission to build five more prototypes with which about 30 selected pilots could be trained. We accepted 60 from the several hundred who had volunteered from the *Luftwaffe* in Friedenthal; especially daring missions would now be possible. Unfortunately, only part of the 500 cubic meters of aviation fuel I requested at the beginning of Summer 1944 was delivered. We could only train the first dozen pilots. The pilots, however, remained in my unit until the end of the war.

The Piloted *V-1... Hanna Reitsch* In Her Own Words

Seldom has so much nonsense, have so many lies and tales been told about an aircraft as in the case of the manned *V-1*. The reasons which led to the manned flights with the *V-1* were never made public. The story began on a day in August 1943, when I met two pilot friends in Berlin. In our discussions we were very much concerned with the worsening situation of our country. We agreed that time was against us. Daily we saw and suffered how the country bled itself to death slowly, how one city after the other was the victim of bombs, how the industry and the traffic were destroyed systematically by the superior air force of the enemies. Supplies were dwindling and death had a horrible harvest among the people. We could also imagine what total defeat meant. We knew that in this war, which had grown into a flood of technology, a turn of the tide would only be possible if we could defeat this monster by its own strength and with the sacrifice of our own lives. Thus, our discussion led to a plan. Germany, we believed, could only be saved from this disastrous situation if a better basis for a faster ending of the war could be negotiated. To achieve this the major strongholds of the enemy and the center of his resistance would have to be destroyed in quick, successive blows, sparing, however, the civil population. This meant destroying power plants, water reservoirs, important industrial sites, and ships in case of an invasion. It was clear to us, that, at this particular time, such a plan could succeed only if there were people, who, even at the risk of their own lives, were ready to attack vital targets to an extent, where their repair and reconstruction were no longer possible. A special technical device would be needed for this. Such missions demanded men and women who were ready to sacrifice themselves with the strong conviction that there was no other means of saving their country. This had nothing to do with false idealism. The plan was only to be realized, of course, if it were proven that a weapon was available which guaranteed success.

This could not be compared with the mission of the Japanese *kamikaze* pilots, whose deaths were a sort of religious sacrifice, and who were content to dive down on an artillery gun on the deck of an enemy warship. This type of mission did not correspond with our European mentality. The plan was carefully weighed and checked by Germany's most capable and renowned scientists, technicians, and leading officers at a Conference before the Academy of Aeronautical Research in the winter of 1943/1944. The plan was

The *USS Essex*. Its single "*kamikaze*" attacker is seen within the circle approaching the aircraft carrier from its stern.

regarded as justified, practicable, and promising success, provided a suitable weapon were available at the right time and did not require a special development.

A manned glided bomb was discussed. First the *Messerschmitt Me 328* was to be used, of which existed one model only. This aircraft was propelled initially by two *Argus As 014* pulse jet engines. It was intended to be used as a fighter, but was canceled by the *RLM* because it did not meet the requirements. I had tested this aircraft at *DFS*-Hoersching near Linz, Austria, for our plan. The requirements for the aircraft to be flown as a glide-bomb were: good vision, good maneuverability, longitudinal stability, and of course, stability. This aircraft (*Me 328*) was tested without an engine. The start was performed piggy-back on the wing of the bomber *Dornier Do 217*.

This gliding-bomb (*Me 328*) seemed to be appropriate for this purpose. A company in Thueringen was ordered to proceed with its production, but this company soon was destroyed by bombs and all specifications were destroyed along with it. Now it was decided to use the *Fi 103R-4* instead of the *Me 328* aircraft.

The unmanned *V-1* was being produced at this time in great numbers and was destined for long-range targets, and was used for the first time on August 16th, 1944. The robot *V-1* was changed into a manned *V-1* within a few days in an underground workshop by the team of *Luthur* and *Gerhard Fieseler* of the *Gerhard Fieseler Werke* (*GFW*). It had the number *Fi 103R-3* and the code-name "*Reichenberg*." All this was kept secret to such an extent that nobody, who was not directly involved, knew about it, not even today. There is no information, neither in German archives nor in museums, maybe because the Fi 103R was never put into service. The manned *V-1* could not be launched with a catapult, like the unmanned *V-1* because of the acceleration of approximately 17 Gs, which are harmful to the human organism. Thus, the manned *V-1* was launched under the right wing of the *Heinkel He 111H* just as it had been done with the unmanned *V-1* configuration, after the launching bases in western Germany had been lost.

Our military test-range at *Erprobungsstelle*-Rechlin rejected my offer to test the manned *V-1* with the argument that this "was a man's job." Thus, I watched from the ground the exciting first launch. After separation from the *He 111H* the pilot ignited the *Argus As 014* pulse jet, which produced an enormous amount of noise for those days. After a few seconds in free flight we observers realized that the pilot had lost control over the airplane. Soon the airplane had disappeared, and a cloud of smoke and detonation indicated disaster. We were very relieved when we learned that the pilot miraculously had been saved, although with bad injuries. Being fully conscious, he could inform us that he had mistakenly actuated the cockpit canopy lock. The canopy had disengaged at 9,000 feet altitude (we on the ground could not see this). By the pressure from the speed of 360 kilometers/hour [224 miles/hour] he momentarily lost consciousness, and so he lost control over the aircraft. A few days later a second manned *V-1* was launched, again with a pilot from the *Erprobungsstelle*-Rechlin test center. He was not a glider-pilot and not used to making spot landings without engine. But the *V-1* had to be landed on a skid, very primitively cushioned. He touched down before the airfield, the *V-1* crashed, and he severely injured his spinal column. *Erprobungsstelle*-Rechlin now wanted to stop the tests, but I continued them, because those accidents were human errors and we had no other weapon to keep our plan going.

Besides myself, six others offered to join in these tests. I wanted to test the first manned *V-1* without an engine in order to test the response of the controls and the stability around its 3 axis, and perhaps improve them before using the *Argus As 014* pulse jet engines. The launching was performed under the wing of a *He 111H* bomber.

It was not exactly enjoyable to fly the manned *V-1*, but by far not so adventurous as it looked to the observer. The glide-angle was very bad. It was very difficult during this first flight to control the ailerons because it was much too sensitive. This meant that a smaller motion of the ailerons led nearly to a roll, and it was therefore difficult to get the *V-1* safely to the ground again. The landing was even more difficult,

The "*kamikaze*" appears to be trailing a plume of fire from the *Essex's* anti aircraft cannons.

because of the landing velocity of about 190 kilometers/hour [118 miles/hour]. After changing the aileron the manned *V-1* could easily be controlled, it fulfilled the normal flight requirements, and could have been flown by any average pilot. But this was not true for the landing, which was extremely difficult and just too great a risk for student pilots. The *V-1's* landing had to be performed, as I told you, on a skid, which was only cushioned by three rubber cushions, because a skid was used only for training. For this reason the manned *V-1* could never be landed with the *Argus As 014* operating, but only with empty tanks and minimum wing loading. This landing could not be demanded from the very great number of pilots who had volunteered for the mission. We decided therefore to build a two-seater manned *V-1* for training purposes.

I tested this aircraft, which turned out to be appropriate for its training purpose. Three or four of the best pilots from all the applicants were to be trained as flight instructors. While they trained the others, these instructors had always to perform the landing themselves. Six pilots, besides me, had flown the manned *V-1*. Two of them had fatal accidents [*Leutnant Starbati* and *Sargent Schenck*], and four were more or less heavily injured. I had the great luck that I stayed accident free all throughout the 10 flights which I performed. Although every flight something exciting and unforeseen happened: Once the *He 111H* intercepted me incidentally shortly after the launching, and it was rather difficult to control the manned *V-1* anymore. On the ground we found that the rear of the fuselage with rudder and elevator was twisted by 30 degrees and that it was a miracle that it had not come off while I performed the difficult landing. Another time the tank-cover had frozen during a test flight with special tanks filled with water, which I performed to about 500 meters (1,640 feet) from 18,000 feet altitude. I was not able to open it, although I tore my skin off my hands. A pilot heavier than I would probably have suffered a spinal injury. It was a miracle that during my two flights with engine on, everything worked well, while my comrades had heavy accidents. The hammering detonations of the *Argus As 014* pulse jet engine loosened the glue of the plywood skin of the wings and caused accidents. One of my comrades [*Heniz Kensche*] was able to eject with a parachute, but only with his utmost strength. Two others could not eject anymore. Probably these detonations were the reason for the many crashes of the unmanned *V-1*. Another reason for the many crashes of the robot *V-1* were the enemy fighters, which only needed to touch the wing tip of the unmanned *V-1* during the flight to induce a crash because the robot had no automatically controlled aileron.

Meanwhile, the invasion in the West had begun and time had overtaken us. Thus, our planned mission had become impossible, and I am glad, because otherwise I would not be alive, unlike many of my comrades who had volunteered for this mission.

The "*kamikaze*" has hit the *USS Essex* on the edge of its port side deck, smashing into fully fueled aircraft.

Operation Suicide

Hanna Reitsch said that the story of the piloted version of the *FZG 76* began about August 1943 in Berlin. She was having lunch at the *Flying Club* with two friends: one was an aviation medical research officer, *Dr.-Med. Theo Benzinger*, who was leader of the Institute for Medical Aeronautics-Rechlin. The second individual at the lunch (unnamed) was a highly experienced glider pilot. During their lunch the topic of conversation turned to Germany's losing war effort. It was agreed that something new must be tried to bring about an end before any more German cities were completely destroyed. The Allies had landed on the beaches of Normandy and were setting about destroying on the *Fi 103 V-1* rocket bomb launch sites. *Reitsch's* plan was to work for a position whereby Germany might obtain a negotiated peace plan. A negotiated peace could only be brought about by weakening the enemy Allied military strength. *Reitsch* believed that this victory could only be obtained in the air, and if enough volunteers were forthcoming then pilots might ride suitable projectiles into the center of Allied targets, destroying them completely and making repair or re-equipment impossible. She claimed that the suicide squadron was not her idea, but that of a friend by the name of *Leutnant Heinrich Lange* from Autumn 1943 and who had been the first individual to place his name on a list of volunteer pilots willing to give up their lives for Germany. *Reitsch* claimed that she and *Lange* had 70 to 80 pilots who had signed suicide pledges to pilot, for example, *Focke-Wulf Fw 190s* on suicide missions. She was not interested in using these pilots for blowing up ammunition trains on the Russian Front as *Hermann Göring* had suggested. That would be a waste of pilots. What she and *Lange* had in mind was hitting Allied heavy naval ships with a 4,000 pound torpedo-like bomb at high speed which would explode beneath the warship. Thus, with the initial 70 to 80 men and woman pilot volunteers, the sinking of heavy ships of war in the English Channel would destroy any attempt at invasion of the European Continent, which many were expecting to happen soon.

Hanna Reitsch may have signed *Lange's* pledge list, but she appears not to have associated herself with the suicide volunteers, but instead wanted to use her name to convince *RLM* officials and high ranking *Nazis* to approve the idea. The conclusion which *Hanna Reitsch* and her two friends came to was to outfit a squadron of piloted projectiles, and in rapid succession, deliver devastating bomb attacks at enemy electric generating stations, waterworks, key production centers, and naval and merchant shipping facilities.

When *Hanna Reitsch* began promoting the plan, she found that many rejected the idea as being out of character of the German people. *Hermann Göring's RLM* rejected the idea as being "fantastic," whereas *Heinrich Himmler* and his *SS* supported *Reitsch* and her idea of using *Fw 190s* as suicide weapons. Then she was able to obtain an interview with *Adolf Hilter* on February 28th, 1944. It appears that *Hilter* was not impressed with *Reitsch's* opinion that Germany's war effort was becoming more desperate each day...so that Germany was likely to lose the war unless drastic measures were taken. Nor did it appear that *Hilter* felt the need for such desperate measures as the formation of suicide squadrons. Nonetheless, *Hitler* apparently gave his permission for *Reitsch* to go ahead with planning, leading to the formation of a suicide squadron. However, its deployment was not to be authorized until he had given the order. *General Korton*, *Luftwaffe* Chief of Staff, was made program director. Also, a *Korton* staff officer by the name of *Oberst Hegl* was to be the liaison person between *Reitsch*, *General Korton*, and *Hermann Göring*. *Oberst Hegl* at this time was commanding *Kampf Geschwader 200* (Combat Air Wing 200). It appears, too, that *Lange*, the individual credited with the idea of a suicide squadron, was made *Staffelkapitän* of a unit known as *5./KG 200*.

The high plume of smoke indicates that the *Essex* has been set afire by the "*kamikaze*" pilot.

Its outward appearing purpose was one of researching unconventional methods of attacking heavily-defended targets, however, its real purpose would have been to carry the *Fi 103R's* to their takeoff point.

Hanna Reitsch's boss at *DFS (Deutsche Forschungsinstitut für Segelfluge*, or German Research Institute For Sailplane Research)*, Professor Walter Georgii,* arranged for a meeting to evaluate the technical and tactical aspects of her plan. In addition to heading *DFS*, *Georgii* was one of the four directors of the powerful German Aeronautical Research Council. He brought together experts in navigation, radio technicians, naval engineers, aircraft designers, explosive and torpedo experts, and finally representatives from the *RLM* and *Luftwaffe* to discuss *Reitsch's* plan for a group of volunteers in a suicide squadron. The outcome of the meeting was that the suicide squadron appeared to be operationally sound, and that a piloted flying bomb should be used with the recommendation that the flying machine be the twin *Argus As 014* pulse-jet powered *Messerschmitt Me 328*. This machine already existed, so it would save valuable construction and development time. The *Me 328* had initially been conceived/designed as a low-wing monoplane fighter. It had lost out in the fighter competition, yet was still being experimented with by *DFS* with two *Argus As 014* pulse-jet powered. *Georgii's* committee of experts had further noted that the *Me 328* could carry up to a 2,000 bomb-torpedo in its fuselage nose, which then could be steered into the water at such an angle that the torpedo would explode directly under the keel of an enemy warship. However, *DFS'* experience with the *Me 328* was unsatisfactory, and the idea for its use as a 2,000 pound bomb carrying machine was dismissed. *Hanna Reitsch* should have known about the *Me 328* because she worked for *DFS* as a test pilot, although there are no reports that she was ever involved in the *Me 328* program.

It was now May 1944, and it appeared that there was no suitable flying machine for the suicide squadron. Another idea was suggested. This involved a standard *Focke-Wulf Fw 190* fighter. The idea was to equip the fighter with a 4,000 bomb, however, it was not known if the *Fw 190* with a nominal weight of 8,600 pounds could lift off a runway carrying a 4,000 bomb, let alone if its tires and landing gear could even support the 2 tons of weight. Even if it could, how long would a runway be for the loaded *Fw 190* to lift off? No one knew. In the end it was decided that a heavily-loaded *Fw 190* would not stand much of a chance in penetrating Allied fighter defenses and reaching their targets.

After a period of time the focus of a suitable machine for the suicide squadron switched from the *Me 328* and *Fw 190* to a piloted version of the *Fi 103*. *Hanna Reitsch* claimed that the switch was due to a shorter operational due date, but mainly on account of the need to carry up to 4,000 pounds of explosives and not the 2,000 a *Me 328* or *Fw 190* could safely carry.

It is not entirely clear how the recommendation came to be that the *FZG 76* be modified to carry both a pilot and a 4,000 pound bomb. As mentioned earlier, *SS-Hauptsturmführer Otto Skorzeny* claims that he was the impetus behind the idea of taking a standard *FZG 76* and adapting it to accommodate a pilot.

Since the flying bomb could hold up to a 4,000 pound explosive charge and its speed was very high, the pilot might have some chance of bailing out before impact. *Hanna Reitsch* said that her boss at *DFS*, *Professor Walter Georgii*, offered the use of his designers and workshop to develop a piloted version of the standard *FZG 76*. In only 14 days *DFS* had converted four *FZG 76s* into piloted versions. These versions were identified by the designations *R-1, R-2, R-3,* and *R-4* based on the code name of "*Reichenberg*." The first three "*Reichenberg*" versions were pilot training machines. The *R-4*, *DFS* recommended, would be carried to the target area beneath an *He 111H* bomber in a similar fashion to the *FZG 76s* launched against England by *Kampfgeschwader 53*.

The end of *Otto Skorzeny* and *Hanna Reitsch's "suicide squadron"* appears to have come about in October 1944. It was at this time that a new *Kommodore* was assigned to *KG 200*, and that person was *Werner Baumbach*. It is believed that he did not favor the use of "*Selbstopfermänner*," or suicide men, but preferred instead unmanned machines carrying high explosives such as the "*mistel*" program...the so-called "father and son" arrangement. This method of attack had been perfected by *Felix Kracht* at *DFS*-Ainring.

As the Allies moved into Germany, discovered were many examples of the "*Reichenberg*." Found were many new-appearing *R-2s*, the two seat training versions at an assembly facility at Pulverhof. All had suffered minor damage to their plastic cockpit canopies, perhaps before they were abandoned. At the Donnerberg factory approximately 125 *R-4s* were found in storage. Quantities of bomb supplies were found, too. It appears that the "*Selbstopfermänner*" program at war's end had not been terminated outright, but that some sort of a training program for the suicide men was getting underway. Perhaps once the victories of the first group of suicide men were seen, a full program would have been initiated. Instead, *Hanna Reitsch* felt that she and her piloted *FZG 76* were looked upon as some sort of stunt instead of the brave, clear-headed, and intelligent Germans who seriously believed after careful thought and calculation, that by sacrificing their own lives they might save many times that number of their fellow countrymen and insure some kind of future for their children:

> I hereby voluntarily apply to be enrolled in the Suicide Group as pilot of a human glider bomb. I fully understand that employment in this capacity will entail by own death.

Reitsch blamed the total failure of her suicide squadron on *Göring*, *Goebbels*, *Milch*, and others, real and imagined, who put obstacles in her way...obstacles greater than she and others could overcome. In addition, she claimed that there were misunderstandings and the clash of personalities among the very people who were working hardest for its success. But in the end, when it appeared that the possibility existed that a training program would be able to go forth...she realized that the decisive moment had been missed (the Allied landing at Normandy June 6th, 1944) and that it was too late. Thus, the lateness of people coming forth

This lone "*kamikaze*" pilot succeeded in killing 15 of the *Essex's* sailors and wounding 44.

to approve the suicide machine and its pilots meant that their was no longer any time left to make a difference. Time had run out on the application of a piloted cruise missile...the *Fi 103 Reichenberg-4. Hanna Reitsch* later said of the whole mess:

> And so did an idea that was born of fervent and holy idealism, only to be misused and mismanaged at every turn by people who never understood how men and women could offer their lives simply for an idea they believed.

Were it not for the grievous damage done to the U.S. Navy's Pacific fleet in late 1944/1945 by the Japanese *Kamikaze* Corps, the German *Fi 103R-4* could be passed off as just another unconventional tactical venture which *Göring*, *Goebbels*, and *Hitler* himself were smart enough to recognize was nothing but foolishness. But in light of the Allies' later experience with the Japanese, it is possible to draw the conclusion that the *Nazi* command failed to realize they were being offered an impressive counter-weapon to seaborne invasion. It is useless, in retrospect, to attempt a reconstruction of what might have happened off Normandy on D-Day, June 6th, 1944, if the *Nazi* command had recognized the potentialities of these volunteers and their piloted flying bomb. Although it is unlikely that *Otto Skorzeny* and *Hanna Reitsch's* suicidists could have defeated the invasion of several hundred thousand men and 5,000 ships, the introduction of such an unconventional tactic, if exploited on the scale later used by the Japanese, would certainly have offered another serious threat to an already difficult amphibious operation. But by three weeks later nearly one million men had arrived, along with 177,000 vehicles and 500,000 tons of supplies; the opportunity to push them back into the sea was long gone.

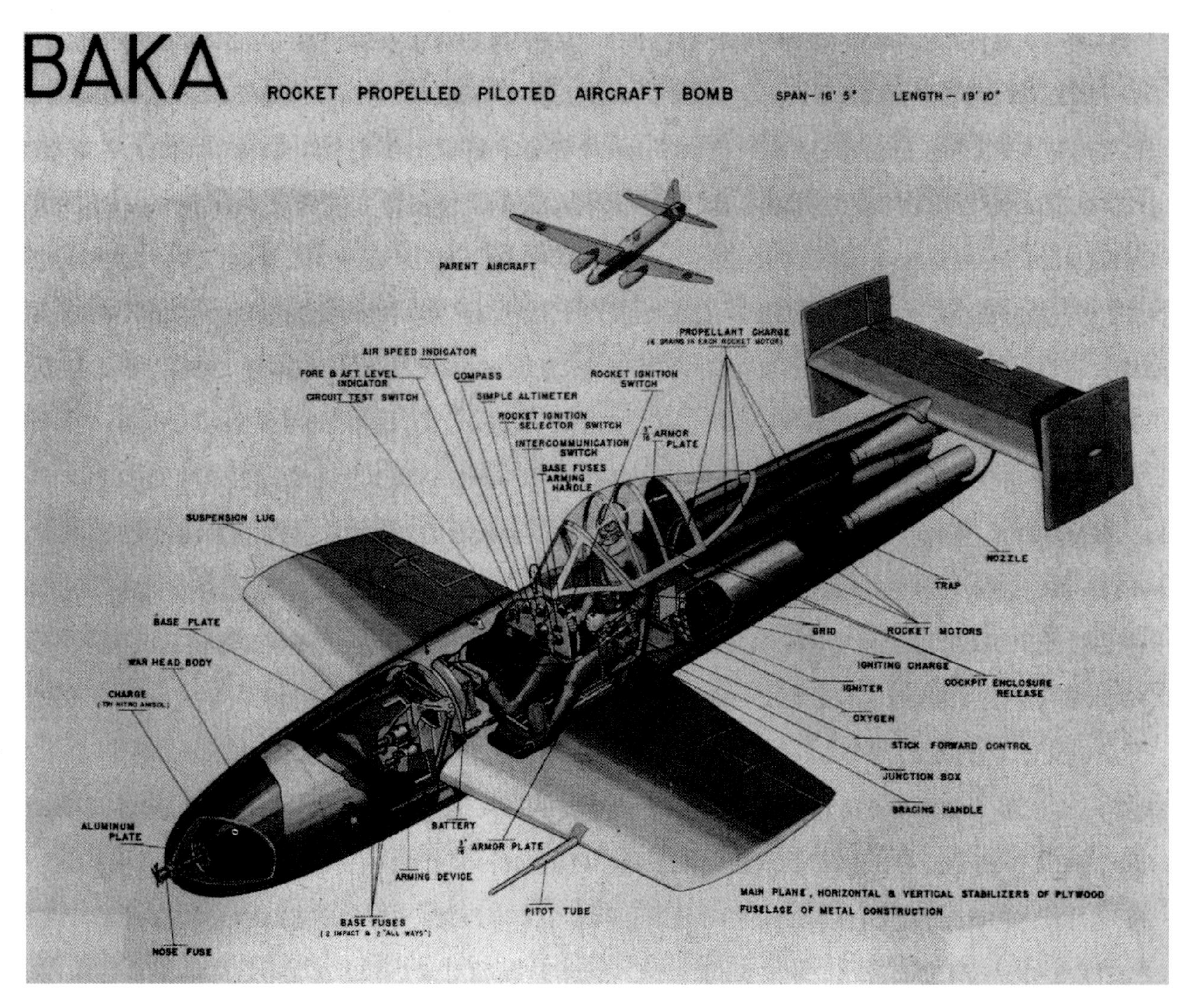

Nicknamed by American intelligence as the *Baka* (fool), this Japanese *MXY*-8 "*Oka 11*" with its 2,645 pound warhead was a specially designed piloted solid fuel rocket missile for suicide attacks on American shipping and invasion forces September 1944 to March 1945.

An "*Oka 11*" has just been released from its carrier aircraft, a *Mitsubishi G4M2E* twin engine medium bomber. The "*Oka 11*" type flying machine first flew in mid 1944.

Above, Right: The "*Oka 11*" was propelled by a group of three *Type 4 Mark 1 Model 20* solid fuel rocket motors giving a total thrust of 1,764 pounds for 9 seconds. At an angle of 50 degrees and its rockets generating full thrust, the "*Oka 11*" could reach a diving speed of 570 miles per hour.

Right: A total of 755 "*Oka 11*" suicide aircraft were produced between September 1944 and March 1945. Three hundred "*Oka 11s*" were shipped to Okinawa to be used against the American invasion forces.

This 4,718 pound loaded weight "*Oka 11*" suicide machine coded "*I-10*" is being collected for American intelligence. It was found on the island of Okinawa on April 6th, 1945.

A poor quality photo of a solid fuel rocket-powered "*Oka 11*" as seen in a Japanese hangar at Katena airfield on the island of Okinawa

An "*Oka 11*" seen in flight. It was carried semi-externally in the bomb bay of a specially adapted *Mitsubishi G4M2E* bomber and was air launched at approximately 200 miles per hour at an altitude of 27,000 feet. It was capable of gliding 50 miles at about 230 miles per hour. During its final approach to the target, the pilot of the "*Oka 11*" fired its solid fuel rockets, and at an angle of 50 degrees, attained a speed of 570 miles per hour to the target.

The "*Oka 11*" had an overall length of 19 feet 8 1/2 inches. It had a wingspan of 16 feet 5 inches.

An "*Oka 11*" coded "*I-18*." It needed no landing gear on its one-way flight to American shipping or invasion forces.

Gerhard Fieseler sitting on the port side fuselage of a bi-wing flying machine. *Fieseler* had a creative genius for designing the appropriate flying machine at the right time. His aircraft factory's wide-ranging work went from the *Fi 103* flying bomb to constructing the prototype twin-engine all-wing *Horten Ho 7* based on their design drawings.

The "*Oka 22*" intended for suicide attacks. It had a 1,320 pound warhead. It did not reach operational service, and about 50 units had been produced by war's end. It differed considerably from its predecessor, especially in its power plant. While the "*Oka 11*" relied on solid fuel rockets, the "*Oka 22*" used a *Tsu-11* four cylinder 110 horsepower motor. This engine drove an axial-flow compressor similar to the Italian *Secondo Campini*. It produced a thrust of 441 pounds, and its maximum forward speed was 276 miles per hour.

Gerhard Fieseler post war.

Gerhard Fieseler's creation...the 26 foot 3 inch long fuselage of the *Fi 103* flying bomb. It was also known as the *FZG 76*. *Skorzeny* and *Reitsch* believed that by rearranging some of its internal components, room could be made for a pilot.

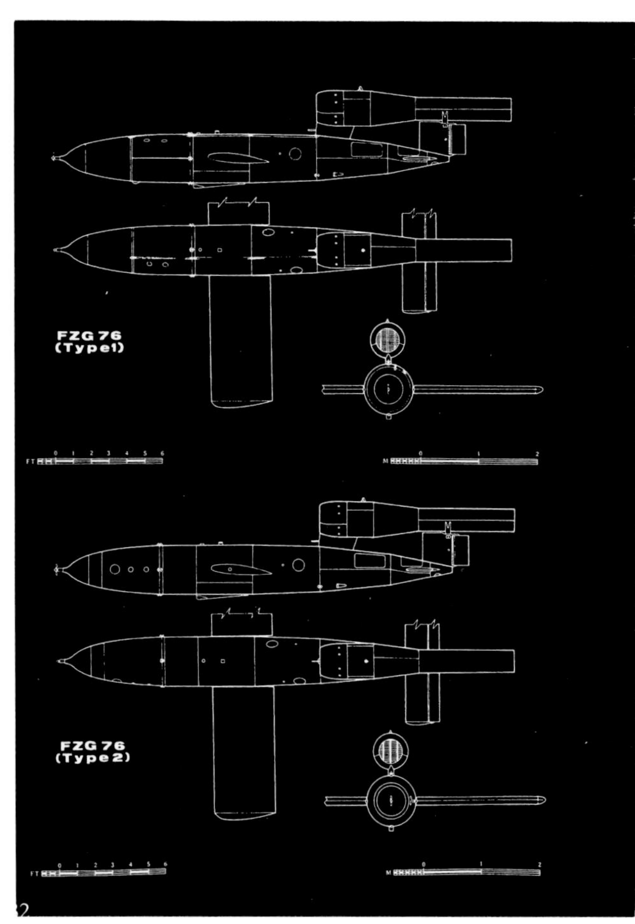

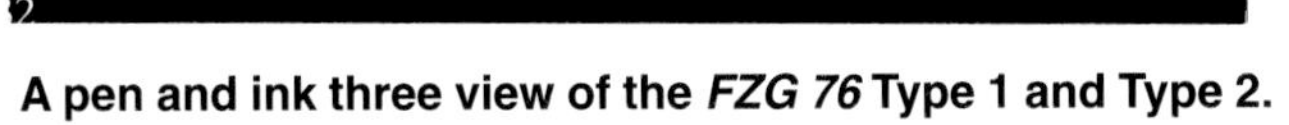

A pen and ink three view of the *FZG 76* Type 1 and Type 2.

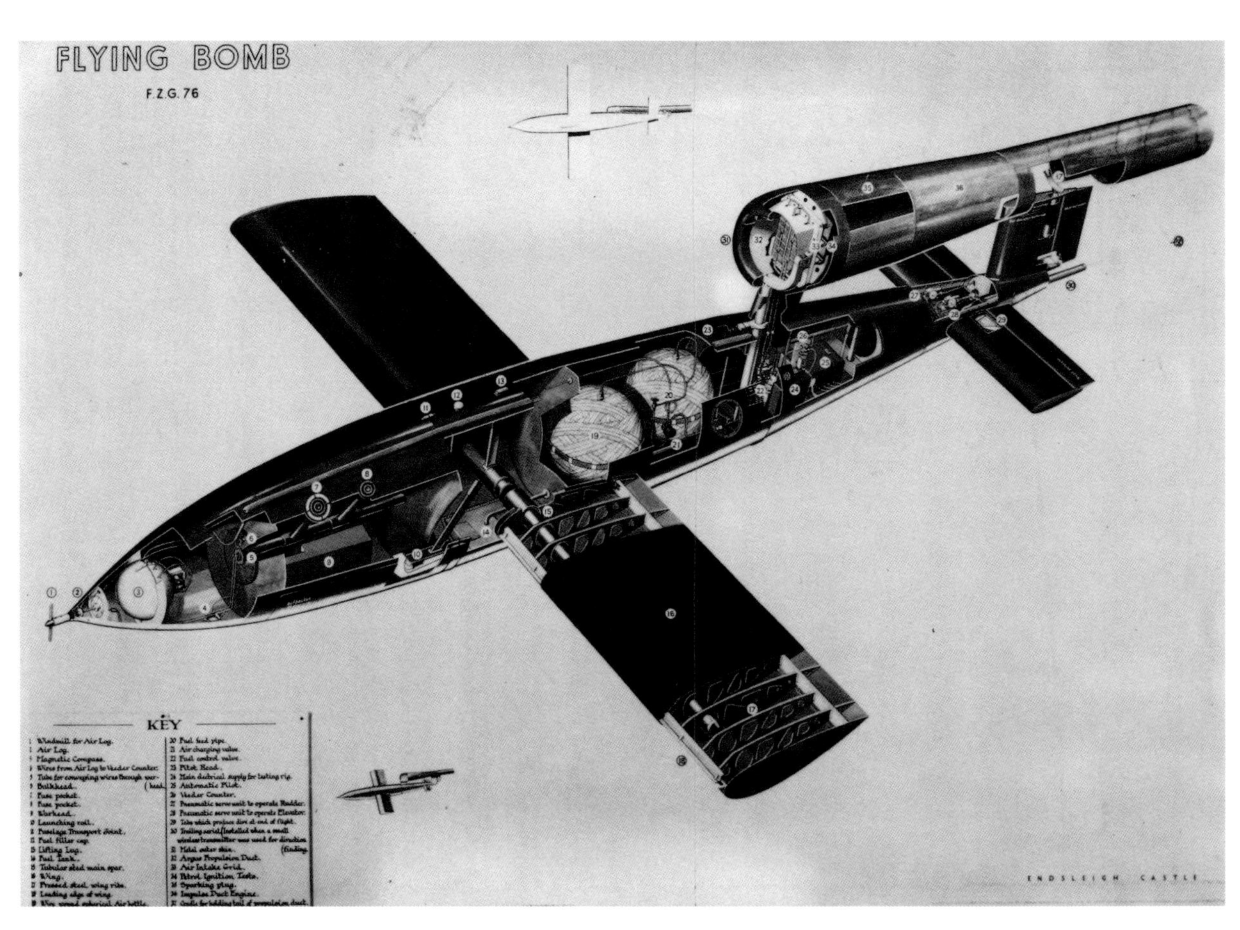

A pen and ink cutaway illustration of an *FZG 76* as seen from its port side.

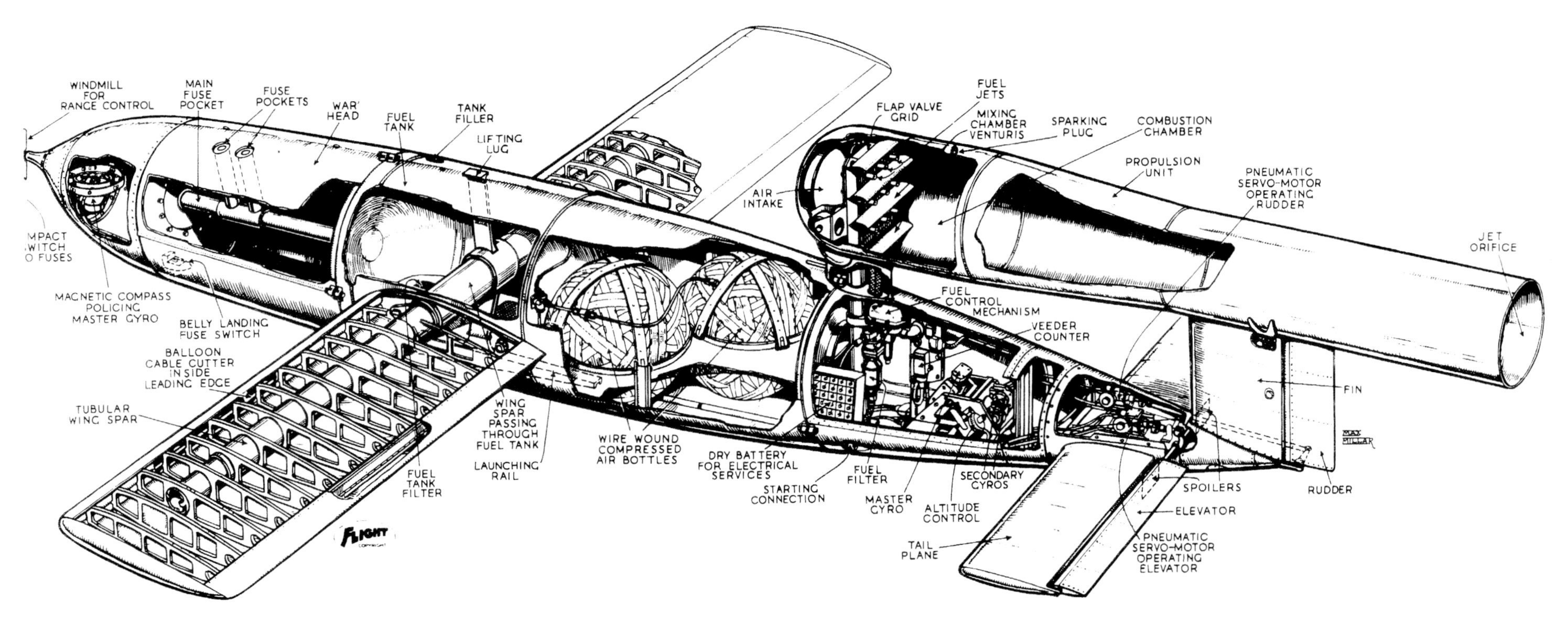

A pen and ink cutaway illustration of an *FZG 76* with its component parts identified.

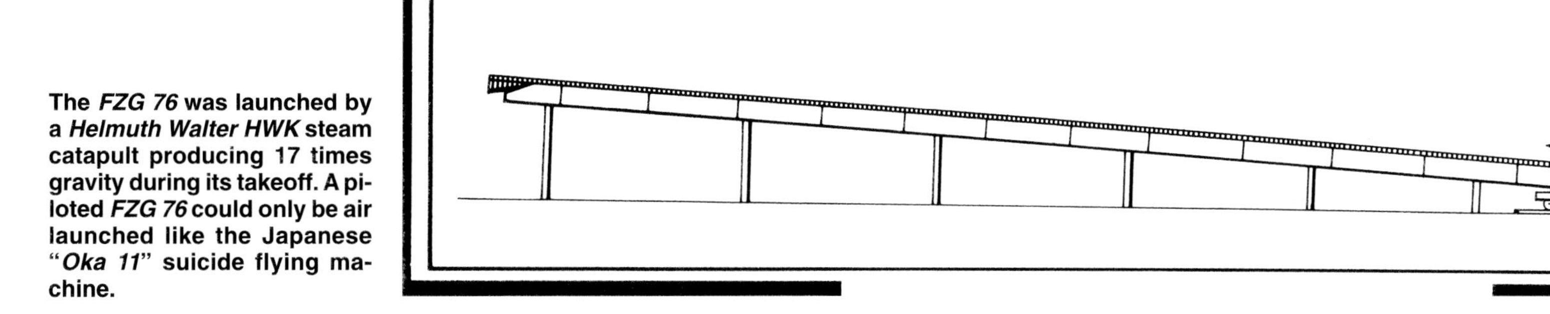

The *FZG 76* was launched by a *Helmuth Walter HWK* steam catapult producing 17 times gravity during its takeoff. A piloted *FZG 76* could only be air launched like the Japanese "*Oka 11*" suicide flying machine.

A restored *FZG 76* firing apparatus complete with launch ramp, a single *FZG 76* mounted on the ramp, and the *Helmuth Walter HWK* steam generating unit.

The *Helmuth Walter HWK* steam generating unit. The flying bomb was attached to the catapult and then high-pressure steam forced the catapult slug up the ramp at 250 miles per hour.

A restored *Fi 103* launch ramp. When the flying bomb reached the end of the ramp it was already speeding along at 250 miles per hour.

The pilotless *Fi 103* with its explosive nose and propelled by a simple pulse-jet engine with no rotating parts is seen here moments after leaving its launch ramp.

A typical launch ramp for the *Fi 103* in France. Destruction seen in the photo was done by the retreating Germans with a few well-placed high explosive devices, especially after the successful Allied invasion of Normandy.

An *Fi 103* flying bomb 2.5 seconds after ignition. Note the flame coming from its pulse-jet. The remaining is steam from the *Helmuth Walter HWK* steam-driven catapult apparatus.

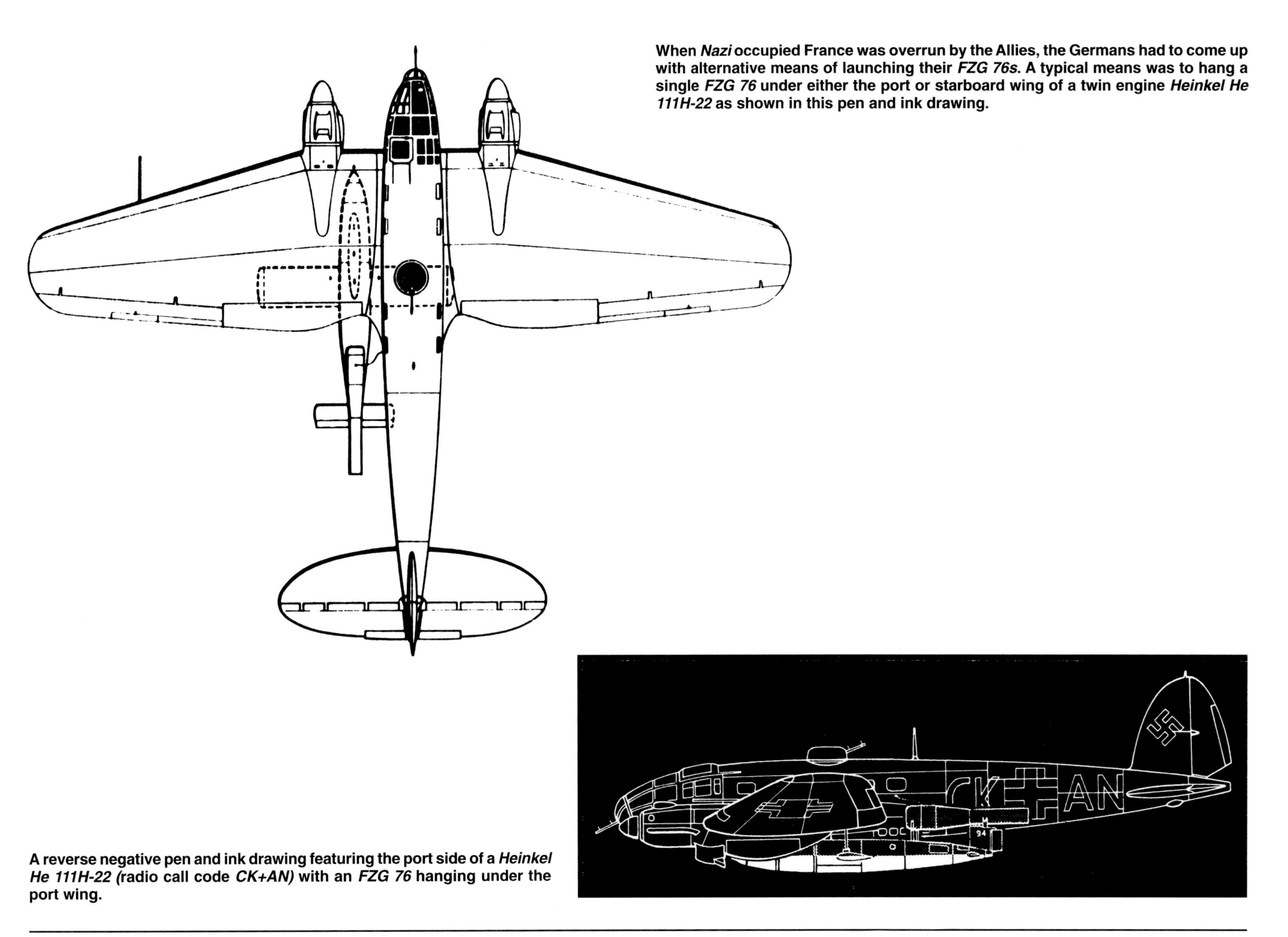

When *Nazi* occupied France was overrun by the Allies, the Germans had to come up with alternative means of launching their *FZG 76s*. A typical means was to hang a single *FZG 76* under either the port or starboard wing of a twin engine *Heinkel He 111H-22* as shown in this pen and ink drawing.

A reverse negative pen and ink drawing featuring the port side of a *Heinkel He 111H-22* (radio call code *CK+AN)* with an *FZG 76* hanging under the port wing.

A nice overall starboard side view of a *Heinkel He 111H* and its external-mounted *V-1* flying bomb.

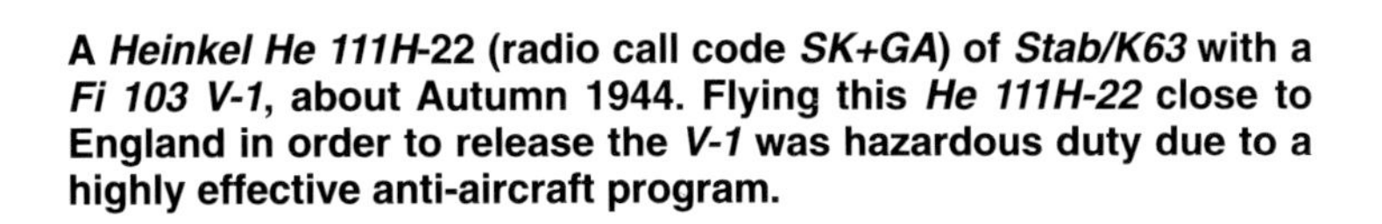

A *Heinkel He 111H*-22 (radio call code *SK+GA*) of *Stab/K63* with a *Fi 103 V-1*, about Autumn 1944. Flying this *He 111H-22* close to England in order to release the *V-1* was hazardous duty due to a highly effective anti-aircraft program.

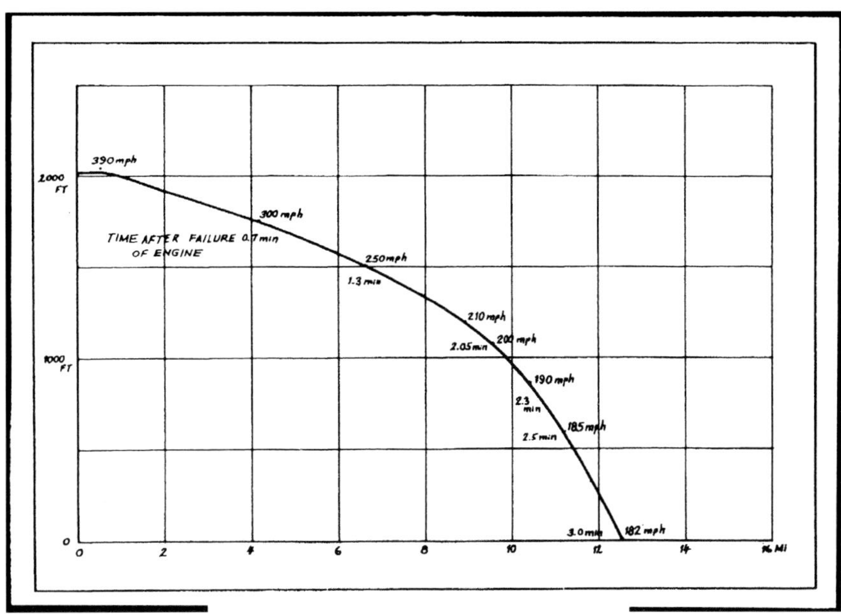

The typical flight path of an air-launched *V-1* flying bomb. When it started its dive at 390 miles per hour, the *V-1* traveled another 13 miles...hitting the ground at 182 miles per hour.

A *V-1* flying blindly over the English countryside at about 2,000 feet and 390 miles per hour.

The pilotless *V-1* would never be able to hit a moving target, such as the *HMS Warspite* (which was hit and severely damaged by a *Fritz X* guided missile). *Skorzeny* and *Reitsch* said that the *V-1* had to be piloted in order to home in on its target.

Hanna Reitsch was one of the top test pilots in Nazi Germany. Here she is shown in the cockpit of a *Focke-Achgelis-Delmenhorst Fa 61*...the first practical helicopter...with mentor *Ernst Udet*.

Hanna Reitsch has the *Focke-Achgelis Fa 61 hovering* above the field inside the *Deutschlandhalle*-Berlin in February 1938.

An intent-looking *Hanna Reitsch* piloting with both hands on the steering yoke.

Hanna Reitsch (March 29th, 1912-August 24th, 1979) 67 years.

FRIDAY, DECEMBER 21, 1945.

Luftwaffe Formed Suicide Squad To Wreck Major Allied Warships

German Woman Pilot Says Unit Organized Late in War Was Expected to Turn Tide but Was Not Used Because of Stupidities

By KATHLEEN M'LAUGHLIN

By Cable to THE NEW YORK TIMES.

OBERURSEL, Germany, Dec. 20—The release today of a further voluminous report resulting from the questioning of Hanna Reitsch, leading woman pilot in Germany who is held at the interrogation center here, reveals that the Luftwaffe possessed voluntary suicide squad fliers who signed a pledge to sacrifice their lives in the later stages of the war to destroy individual units of the Allied fleets, expecting thus to turn the tide of conflict.

That it never was utilized Fraeulein Reitsch credits to the stupidities of Adolf Hitler, Hermann Goering and Heinrich Himmler.

Another section of the report details Fraeulein Reitsch's vitriolic accusations against Goering, whose incompetence and egomania, she asserts, cost hundreds of thousands of lives through his falsification of German air production figures and distortions of the actual status of the Luftwaffe in his conferences with the Fuehrer. These misrepresentations, she declares, prolonged the war for months after the clearer minds among the German leaders recognized it was irrevocably lost.

Wrecked Organization

Such mismanagement and dilatory tactics ultimately completely wrecked the organization, the woman flier asserts, while revealing simultaneously her vigorous machinations in obtaining the removal of certain officials who sought to divert "the idealistic spirit" of the death squadron to their own selfish aggrandizement with entire disregard to its original purpose.

After the time element forced a shift in plans from the ME 328 and FW 190 to the V-1, frequent fatalities occurred in testing the craft to determine its maneuverability and potentialities, Fraeulein Reitsch relates. The Luftwaffe testing divison ignored her offers to act as test pilot and two of its own fliers assigned to the job each crashed twice, receiving serious injuries, after which she was called in and allowed to continue the trials.

FZG 76 (Type 1)

FZG 76 (Type 2)

Biggest Warships Targets

As one of the two individuals most responsible for the suicide squadron, she explains that it was organized with the objective of allotting each member a specific target, the list comprising the mightiest warships massed against Germany. Each ship was to be utterly destroyed by a 4,000-pound bomb that would have been automatically directed to the keel after the pilot had hurtled toward the water at terrific speed and the torpedo type of bomb fitted with a time fuse had exploded beneath the keel.

She herself, as test pilot of the V-1 projectiles being adapted for the operation, attained a speed above 528 miles an hour, she says, as indicative of the momentum that was to have resulted in the disintegration of the weapon on impact with the water and thus carried the volunteers to their deaths with typical Teutonic violence.

Picturing the basis of the plan as the conviction of a small group of idealists that a sudden and devastating blow against the Allies would effectively destroy any attempt at invasion of the Continent that many realized was inevitable, Fraeulein Reitsch recites that the actual organization came into being in the autumn of 1943 under the leadership of First Lieut. Heinrich Lange. Her name headed the list of those signing the pledge drawn up, although she never formally joined the group, preferring to exercise her great influence in its behalf among topflight circles to which she had entrée.

Dismissed as Fantastic

After the German Air Ministry dismissed the proposal as fantastic, Fraeulein Reitsch, Lange and Dr. Theo Benzinger, leader of the Institute for Medical Aeronautics at Rechlin, arranged a meeting to evaluate the technical and tactical capabilities. It was called by Dr. Walter Goergil, director of the German Aeronautical Research Council, and included as participants navigation and radio technicians, ship engineers and naval experts, experienced airplane designers and constructors, explosive and torpedo experts and representatives of the Luftwaffe fighter and bomber commands and the office of the Air Surgeon.

This session came to the conclusion that the plan was operationally sound; that a manned flying bomb should be used, with the recommendation that it be the Messershmitt 328, which already existed, thus saving valuable construction development time; and that a 2,000-pound bomb-torpedo should be installed in the nose of the plane that would be steered into the water at such an angle that the torpedo would explode directly under the keel.

The pressure of time as the war approached the critical stage and opposition and sabotage among air and Government executives developed forced an alteration of the plans to the projected use of the V-1 and to double the size of the explosive it would carry.

Many pages of the report cite Fraeulein Reitsch's frustrations in promoting the plan in February, 1944, and those who rejected the idea as out of character with the German people; with Himmler, who spiked her efforts to win acceptances for the Focke-Wulf 190 as a suicide craft through relaying that development to Hitler; and the tardy efforts by Goering to utilize most of the seventy to eighty signers of the suicide pledge by sending them on such insignificant missions as blowing up ammunition trains on the Russian front, which could not possibly justify their certain deaths.

She had made ten flights when she was injured in an air attack on Berlin, and Lieutenant Starbati took up the experimental work. He was killed when a V-1 crashed, and was replaced by Sergeant Schenck, who also was killed in a similar accident. Then Heniz Kensche, in charge of the venture, tried the operation and narrowly escaped death when his craft fell out of control as a result of excessive vibration at a low altitude. It was subsequently established with great difficulty that this factor was due to the full power action of the Argus-Schmitt power units and was causing at least 30 per cent of the unmanned V-1's launched against England to crash through mechanical failure before reaching the target.

Denounces Goering

Fraeulein Reitsch was almost hysterical in her denunciation of Goering as the malevolent perpetrator of much deception with other Nazi executives, and blindly refusing actual aircraft production statistics, while insisting that possibly as a result of his drug habit he was making incredible statements about air force equipment and strength. His "Caesar complex" was so intense, she claims, that he dismissed in disgrace any technician or administrator who fought to get the facts before the proper authorities.

She was particularly venomous when referring to the superior abilities of Gen. Oberst Ritter von Greim, commanding general of Luftlotte Six, whom Hitler appointed as successor to Goering as Luftwaffe chief in the final days of the war. His honor as a soldier inspired his suicide at Salzburg last May, she claims, in preference to telling the world the truth about his predecessor, who, he felt, had contributed vastly to the catastrophe Germany suffered.

New York Times article "*Luftwaffe* Formed Suicide Squad To Wreck Major Allied Warships."

Luftwaffe leader *Hermann Göring* said no to *Hanna Reitsch's* "Suicide Squadron." Postwar she blamed Germany's defeat on "the fat one," whom she had come to hate. He was not around to answer *Reitsch*, having committed suicide in his prison cell at Spandau moments before he was to be executed. This photo shows how *Göring* looked prior to his suicide. The "fat one" was now much thinner than before.

Hanna Reitsch's second most hated *Nazi* personality was *Joseph Goebbels*, Minister of Propaganda. He, too, did not support her efforts for a piloted *FZG 76* or her ideas for a suicide squadron. *Goebbels* took his own life and that of his wife and their six children in *Hitler's* bunker on May 1st, 1945.

The 5 foot 1 inch *Hanna Reitsch* deplaning the "*Habicht*," or Hawk sailplane. She weighed about 100 pounds.

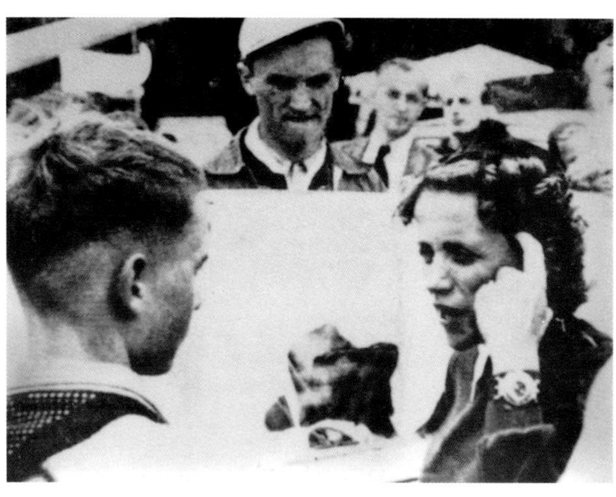

Hanna Reitsch chewing out a young male sailplaner. With her finger touching her temple, she is graphically showing this boy that he is some sort of dumb head. Not a nice gesture.

Right: *Professor Walter Georgii*, head of the *DFS*-Ainring and *Hanna Reitsch's* supportive and protective boss.

Hanna Reitsch with *Heinz Scheidhauer*, former sailplane test pilot for the *Horten* brothers. She never married or had children. Two of her significant others included *General Ernst Udet*, chief of *Luftwaffe* planning and *General Robert Ritter von Greim*, the last chief of the *Luftwaffe*. Both men committed suicide: *Udet* on November 17th, 1941, and *von Griem* on May 24th, 1945.

Opposite, Top Left: *Ernst Udet* (center) and *Oskar Ursinus* (far right), father of the *Rhön/Wasserkuppe*. *Udet* was *Reitsch's* good friend, mentor, and perhaps lover. He committed suicide November 1941 due to criticism that he was responsible for Germany's failure in the Battle for Britain, the Heinkel He 177 bomber, and so on.

Opposite, Bottom Left: *General Robert Ritter von Greim*...the last chief of the *Luftwaffe*. He is also considered *Hanna Reitsch's* good friend, mentor, and perhaps lover. She flew *von Greim* on the improbable round trip to *Hitler's* Berlin bunker. He and *Reitsch* were *Hitler's* last guests before committing suicide in the bunker. *Von Griem*, too, would commit suicide shortly after being arrested by the Allies in May 1945.

Left: *Hanna Reitsch* receiving the Iron Cross, 2nd class from *Hitler* on March 28th, 1941, as *Hermann Göring* looks on with great amusement. She was awarded the Iron Cross for her contribution as a first-rate test pilot.

The ill-fated *Messerschmitt Me 328* interceptor shown in pen and ink. *Hanna Reitsch* was hoping to have this *Argus As 014* pulse jet powered flying machine for use in attacking Allied ships of war.

The *Messerschmitt Me 328* had been given over to *DFS*-Ainring for development. This is where *Hanna Reitsch* first became acquainted with this flying machine. *DFS* was not able to turn the machine into an interceptor due to the excess noise and vibration caused by its twin *Argus As 014* pulse jet engines. Project canceled.

The *Messerschmitt Me 328* would have been carried up to its twin pulse-jet's operating environment on the back of a *Dornier Do 217.* Flight tested, but abandoned as a potential "kamikaze" machine.

A wind tunnel model of the *Messerschmitt Me 328* interceptor. In this version its twin *Argus As 014* pulse-jets were to be mounted on the fuselage sides aft the wing's trailing edge. Project abandoned due to excessive noise and vibrations.

Consideration was given to arming the 8,000 pound *Focke-Wulf Fw 190* with a 4,000 pound bomb. *Hanna Reitsch* and others would fly the *Fw 190* as a "*kamikaze*" machine. Could an *Fw 190* even lift off with 4,000 pounds of extra weight? Never tested. Idea abandoned.

The "*mistel*" program involved two aircraft...the bottom carrying the explosive device. The upper flying machine (*Focke-Wulf Fw 190*) was piloted and put the aircraft with the explosive charge on a flight line to the target. However, this combination would not have been effective in slowing or even stopping the Allied invasion...the machinery was too slow and could be easily shot out of the sky before it had a chance to target military shipping.

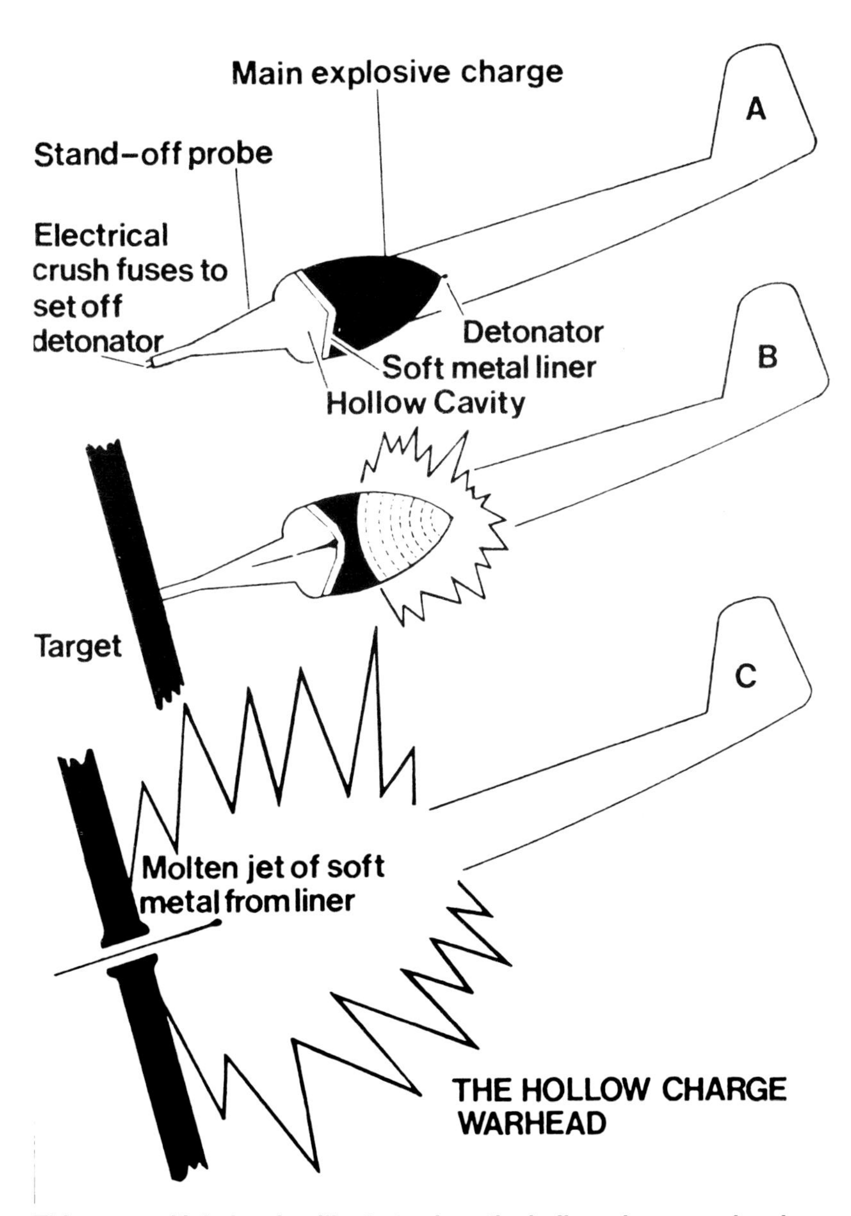

This pen and ink drawing illustrates how the hollow-charge warhead applied to the "*mistel*" program penetrated thick steel.

Hitler **(far left) and** ***Mussolini*** **(center with arms folded) in the mid 1930s.**

Liberation day for "*El Duce*" September 12th, 1943. *Skorzeny* (left) and *Mussolini* (center).

Feldmarschall Erhard Milch. 1892-1972

The rear port side view of an *FZG 76.* This pilotless flying machine was already pretty much packed with components. *Skorzeny* and his colleagues had a challenge, especially when it came to fitting a pilot into this cramped space.

Feldmarschall Milch, according to *Skorzeny*, gave him permission to convert up to three *FZG 76* flying bombs to piloted versions. Yet *Milch* was not so liberal to *Hanna Reitsch.*

A port side nose view of an *FZG 76* of the exact same type *Skorzeny* set about converting into a piloted version. The job was accomplished in a matter of weeks rather than a matter of months. *Feldmarschell Milch* was highly impressed with *Skorzeny's* organizational skills.

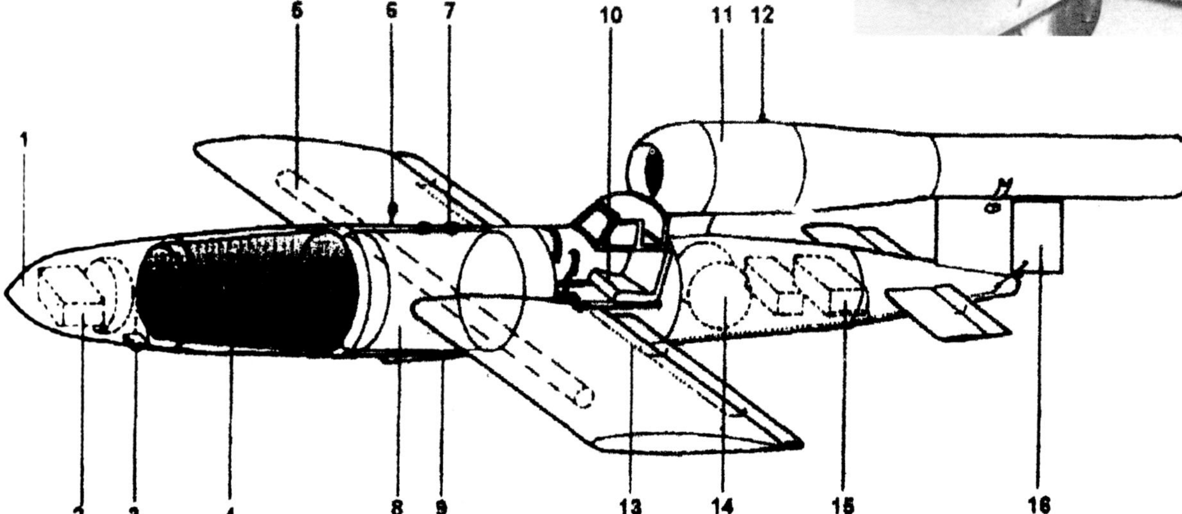

A pen and ink drawing featuring how *Skorzeny* modified the basic *FZG 76* into a manned version.

01 - streamlined nose shell;
02 -
03 -
04 - explosive warhead;
05 -
06 - tubular steel spar;
07 -
08 -
09 - *Walter HWK* catapult attachment;
10 - pilot's cockpit;
11 - *Argus As 014* pulse-jet engine;
12 - single sparkplug;
13 - ailerons;
14 -
15 -
16 -
17 -
18 - rudder:

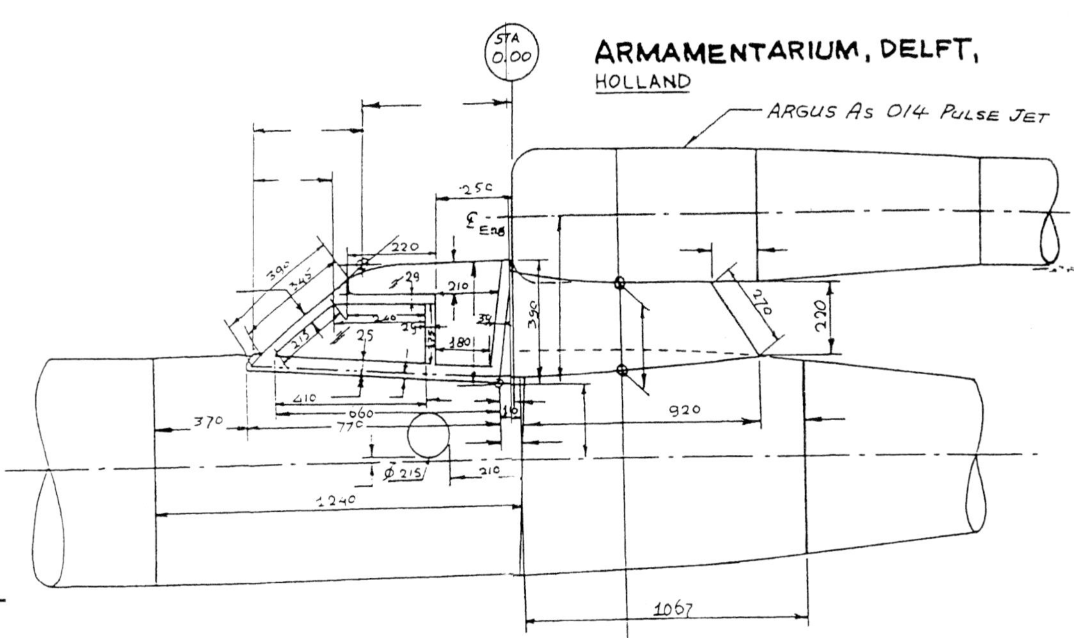

A pen and ink drawing featuring the port side measurements for the cockpit canopy on the manned version of the *FZG 76*.

One of the *V-1* converted into a piloted *V-1*. The round opening forward the cockpit canopy is where the flying machine's tubular spar passed through to the opposite side.

This overhead view of a *V-1* features the tubular metal spar. A ground crew is attaching the port and starboard side wing panels by sliding them on to the metal pipe wing spar.

A close-up view of the plexiglass panels in the cockpit windscreen. Notice, too, how the cockpit canopy butts right up to the *Argus As 014's* air intake. It does not appear to give a lot of room for a pilot struggling to leave the flying machine and not get caught up in the *Argus's* air intake.

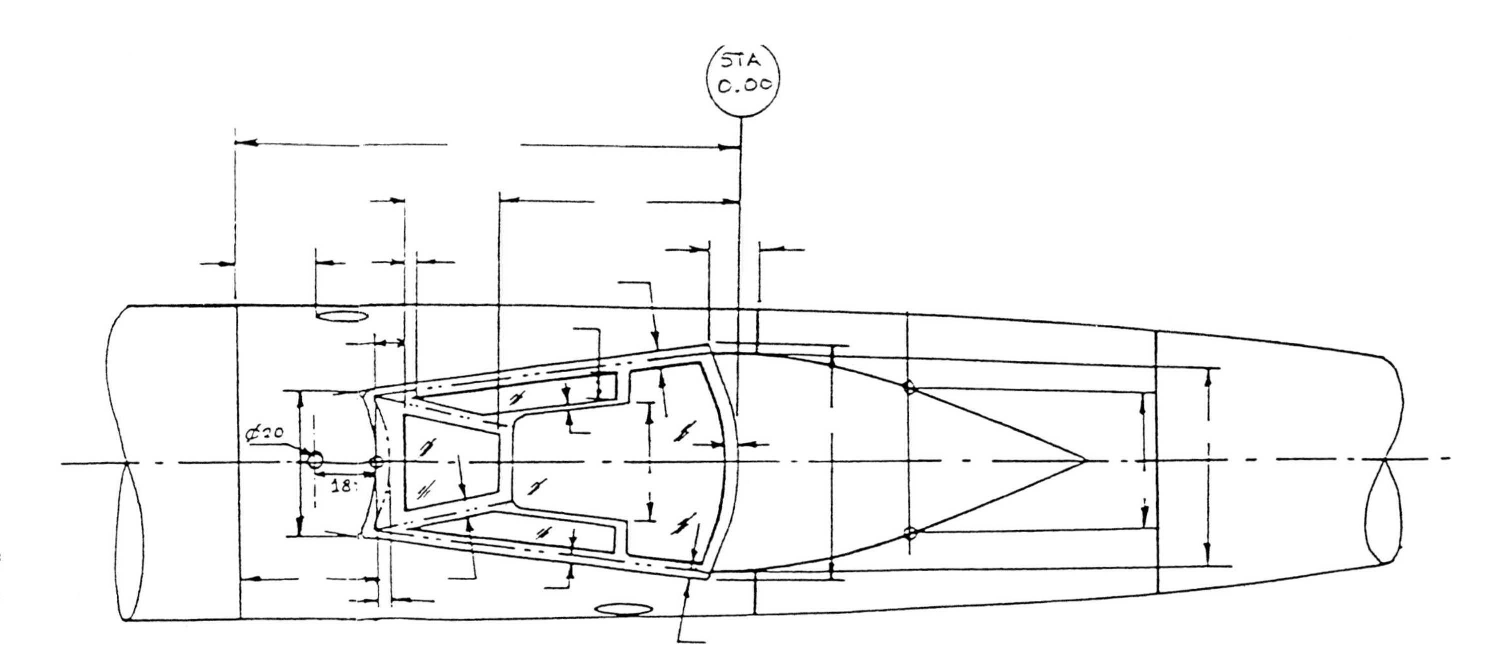

A full top view, with measurements, of the cockpit windscreen, top panels, and rear cockpit fuselage fairing.

A piloted *V-1* seen from below and featuring its cockpit windscreen. Skorzeny's design team produced a very pleasing aerodynamic cockpit canopy. Digital image by ***Mario Merino***.

Left: The standard *FZG 76* as it appears looking aft down its long nose to the *Argus As 014* pulse-jet engine.

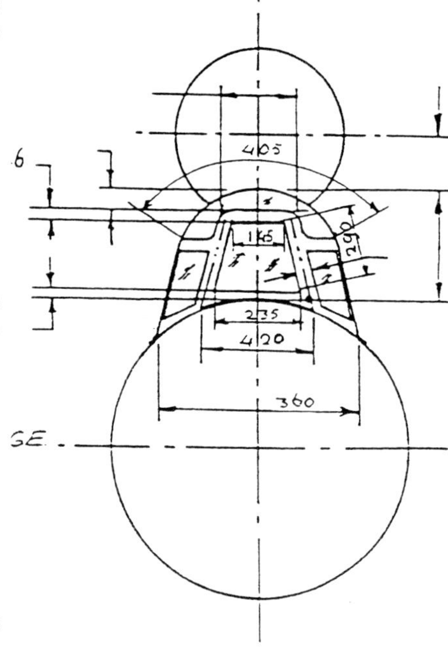

Right: Pen and ink measurements for the proposed cockpit canopy on the converted *FZG 76*.

Below: The real thing...the piloted *V-1's* direct front-on view of its cockpit canopy.

A piloted *V-1* in captivity post war and seen from its port side.

A piloted *V-1* post war and seen from its port rear side.

***Feldmarschall Milch* believed that the standard *FZG 76* could be converted into a piloted version given enough time. He was highly supportive in the hardware aspect of the "*kamikaze*" machine, but he was less clear regarding a program whereby *Luftwaffe* pilots would sacrifice their life in order to destroy Allied shipping.**

Skorzeny (left) and his former longtime adjutant *Radl* photographed in a POW stockade at Darnstadt.

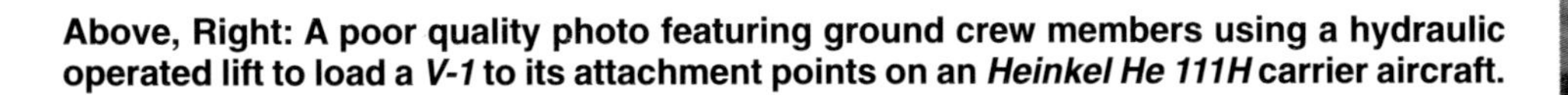

Above, Right: A poor quality photo featuring ground crew members using a hydraulic operated lift to load a *V-1* to its attachment points on an *Heinkel He 111H* carrier aircraft.

Right: The *V-1* is secured ready for take off under the starboard wing of a *Heinkel He 111H*.

One of the three *Skorzeny* converted piloted *V-1s* being carried to altitude under the starboard wing of a *Heinkel He 111H*.

An artist's rendering of the *Skorzeny* converted piloted *V-1s* shown to *Feldmarschall Milch*.

One of three *Skorzeny* prototype piloted *V-1s* in flight.

Dust and smoke tell that the prototype piloted *V-1* came down fast and hard upon landing at Peenemünde. The pilot was severely injured.

Hanna Reitsch, disobeying orders, seen moments before giving the piloted *V-1* a test flight. She is shown at Peenemünde.

An *Fi 103R-4* belonging to the Lashendon Air Warfare Museum at Headcorn in Kent, England, 1992. It is undergoing restoration by *Trevor Matthews* and colleagues. *Matthews* believes that this piloted flying bomb was one of several which *Hanna Reitsch* test flew at Peenemünde, where this example was found. *Reitsch* said that she made as many as 10 test flights in piloted *V-1s*...all out of Peenemünde. Courtesy: *Bob Ogley*, "Doodlebugs and Rockets," Froglets Publications, Ltd., Brasted Chart, Westerham, Kent, England.

A poor quality photo of *SS Komando Otto Skorzeny* and featuring his decorations. He was the prime mover on getting FZG 76s converted into piloted versions. He credited his success to *Feldmarschall Milch.*

REICHENBERG IIA

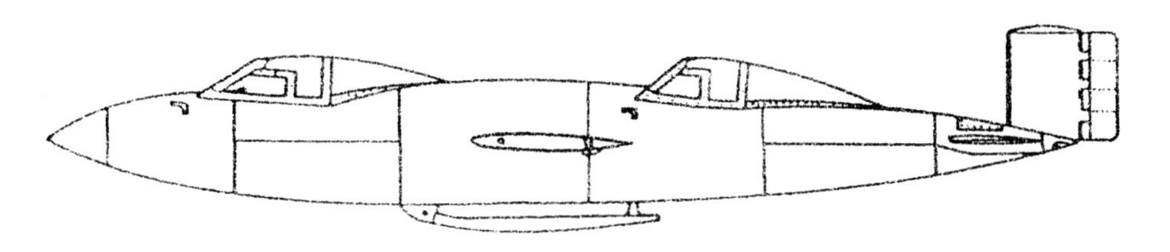

REICHENBERG II B

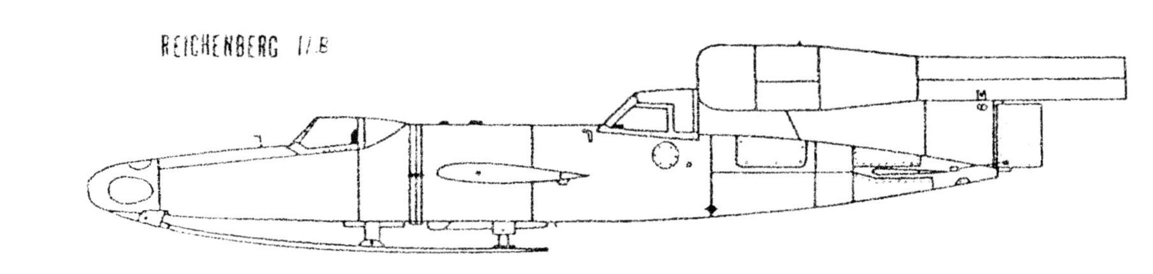

REICHENBERG III

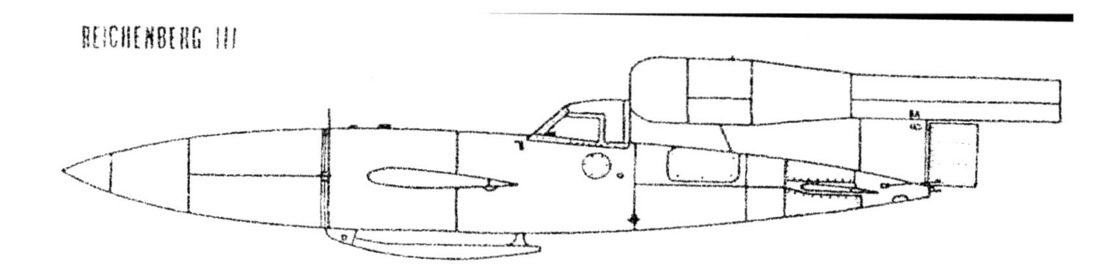

Confusion here. *Otto Skorzeny* implies that he was the individual who had the three piloted *V-1* trainers designed...the *Reichenberg #1*, *#2*, *#3*, and *#4*. Perhaps. On the other hand, *Hanna Reitsch* writes that the piloted *V-1* trainers were built at *DFS*-Ainring through the efforts of *Professor Walter Georgii*, head of *DFS*-Ainring. Perhaps.

REICHENBERG IV

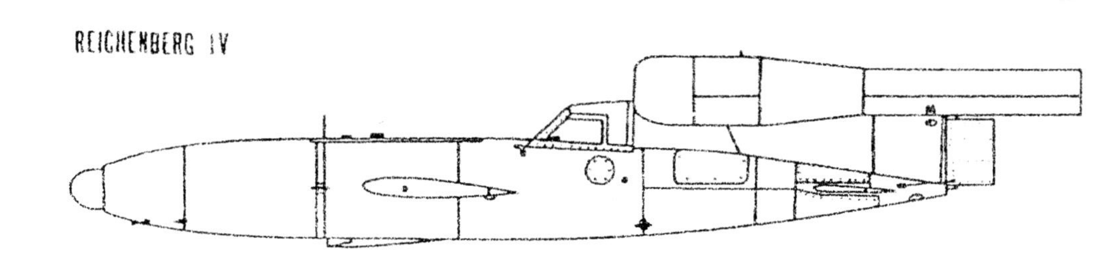

WEIGHTS AND PERFORMANCE STATISTICS									
Type		Fi 103 A-1	Fi 103 A-1/Re 1	Fi 103 A-1/Re 2	Fi 103 A-1/Re 3	Fi 103 A-1/Re 4	Fi 103 B-1	JB-2	LTV-N-2
Role		Missile	Extd. Range Missile	Towed Missile	Missile Trainer	Piloted Missile	Missile	U.S. Army Missile	U.S. Navy Missile
Seating		—	—	—	2	1	—	—	—
Engine		Argus 109-014	Argus 109-014	—	Argus 109-014	Argus 109-014	Porsche 109-005	Ford PJ-31-F-1	Ford PJ-31-F-1
Thrust	km-km/h-kp (lb) km-km/h-kp km-km/h-kp	0-0-366 (807) 0-600-325 (716) 3-600-254 (560)	0-0-366 (807) 0-600-325 (716) 3-600-254 (560)	—	0-0-366 (807) 0-600-325 (716) 3-600-254 (560)	0-0-366 (807) 0-600-325 (716) 3-600-254 (560)	0-0-500 (1,102) 0-644-338 (750)	0-0-405 (900)	0-0-405 (900)
Fuel	Liter (US gal.)	665(176)75 octane	680 (180) VK2	—	680 (180)	680 (180)		680 (180)	680 (180)
Engine weight	kg (lb)	153 (337)	153 (337)	—	153 (337)	153 (337)	199 (441)		
Empty weight	kg (lb)	839 (1,864)	1,204 (2,675)						
Fuel weight	kg (lb)	500 (1,102)	513 (1,131)	—					
Crew weight	kg (lb)	—	—	—	202 (448)	101 (224)	—	—	—
Warhead weight	kg (lb)	830 (1,830)	450 (1,000)	830 (1,830)	—	810 (1,800)	842 (1,870)	945 (2,100)	945 (2,100)
Equipped weight	kg (lb)	2,152 (4,796)	2,163 (4,806)					2,261 (5,025)	2,260 (5,023)
Fuel consumption	L-kp (gal-lbst)	2.59-272 (.68-605)	2.59-272 (.68-605)	—	2.59-272 (.68-605)	2.59-272 (.68-605)			
Optimum range	km (mi)	238 (149)	375 (233)				700 (435)	242 (150)	242 (150)
Service ceiling	m (ft)	2.625 (8,840)					3936 (13,120)	1800 (6,000)	1800 (6,000)
Optimum cruise speed	km/h (mph)	580 (360)	628 (390)				650 (404)	644 (400)	644 (400)
Maximum speed	km/h (mph)	644 (400)	773 (480)				900 (558)	708 (440)	708 (440)
Rate of climb	m-min (ft-min)			—	—	—		300 (1,000)	300 (1,000)
Speed at launch	km/h (mph)	400 (248)	400 (248)	—	—	—		354 (220)	354 (220)
Launch time	sec	1	1				1	1	1
Duration of flight	min	25	40					25	25
Warhead		Amatol-39 (CODE Nr. 52A)	Amatol-39 (CODE Nr. 52A)	Amatol-39 (CODE Nr. 52A)	—		SC 800	T-8 Light case bomb	T-8 Light case bomb

[1] RLM drawing wing span 5370 mm. Manufacturer's drawings list 5382 mm (17 ft-6 3/4 in) for wooden wings and 5736 mm (18 ft-10 in) for steel wings. Other wings were tested including a tapered wing of 4870 mm (15 ft-11 5/8 in) span with a root chord of 1340 mm (4 ft-0 in) and tip chord of 874 mm (2 ft-8 in).
[2] Approximate dimension.

SPECIFICATIONS									
Type		Fi 103 A-1	Fi 103 A-1/Re 1	Fi 103 A-1/Re 2	Fi 103 A-1/Re 3	Fi 103 A-1/Re 4	Fi 103 B-1	JB-2	LTV-N-2
Wing span	mm (ft-in)	5370 (17-6¼)[1]	5370 (17-6¼)	6850 (22-4⅞)	5720 (18-9)	5720 (18-9)	5300 (17-4⅝)	5392 (17-8⅛)	5392 (17-8⅛)
Length overall	mm (ft-in)	8325 (27-3¾)	8509 (27-11)	7405 (24-3⅜)	8929 (29-9)[2]	8380 (27-2)		8262 (27-1 1/16)	8262 (27-1 1/16)
Height	mm (ft-in)	1423 (4-8)	1423 (4-8)		1423 (4-8)	1423 (4-8)		1423 (4-8)	1423 (4-8)
Stabilizer span	mm (ft-in)	2055 (6-8⅞)	2055 (6-8⅞)	—	2055 (6-8⅞)	2055 (6-8⅞)			
Fuselage length	mm (ft-in)	7405 (24-3⅜)	7772 (25-6)	7405 (24-3⅜)	8323 (27-8)[2]	7780 (25-2)			
Maximum fuselage diameter	mm (ft-in)	840 (2-9)	840 (2-9)	840 (2-9)	840 (2-9)	840 (2-9)	840 (2-9)	858 (2-9¾)	858 (2-9¾)
Engine length	mm (ft-in)	3660 (12-0⅛)	3660 (12-0⅛)	—	3660 (12-0⅛)	3660 (12-0⅛)	2853 (9-4¼)	3543 (11-7½)	3543 (11-7½)
Maximum engine diameter	mm (ft-in)	565 (1-10¼)	565 (1-10¼)	—	565 (1-10¼)	565 (1-10¼)	650 (2-1⅝)		
Engine orifice diameter	mm (ft-in)	390 (1-3⅜)	390 (1-3⅜)	—	390 (1-3⅜)	390 (1-3⅜)		390 (1-3⅜)	390 (1-3⅜)
Wing chord	mm (ft-in)	1050 (3-5¼)	1050 (3-5¼)		1296 (4-3)	1296 (4-3)			
Wing area	m^2 (ft^2)							1.71 (60.5)	1.71 (60.5)

Weights, Performance Statistics, and Specifications of the *Fieseler Fi 103* - All Versions

An artist's rendering of the tandem-seat *Fi 103 R-2A* trainer version. This machine was an unpowered sailplane.

Two tandem-seat *Fi 103 R-2Bs* found postwar. There are differences between the *R-2A* and *R-2B*. The latter, seen in the photo, have a more streamlined front seat, plus they carried an operational *Argus As 014* pulse jet engine. The machines shown have had their front cockpit canopy smashed in by retreating Luftwaffe personnel. Notice that the wooden landing skid begins just aft the fuselage nose.

An artist's rendering of the *Fi 103 R-3* single seat trainer with an operational *Argus* As *014* pulse jet engine. However, this machine was used with and without its pulse jet operating.

A very rare photo of an *Fi 103 R-3* in flight at high altitude. No place or date given.

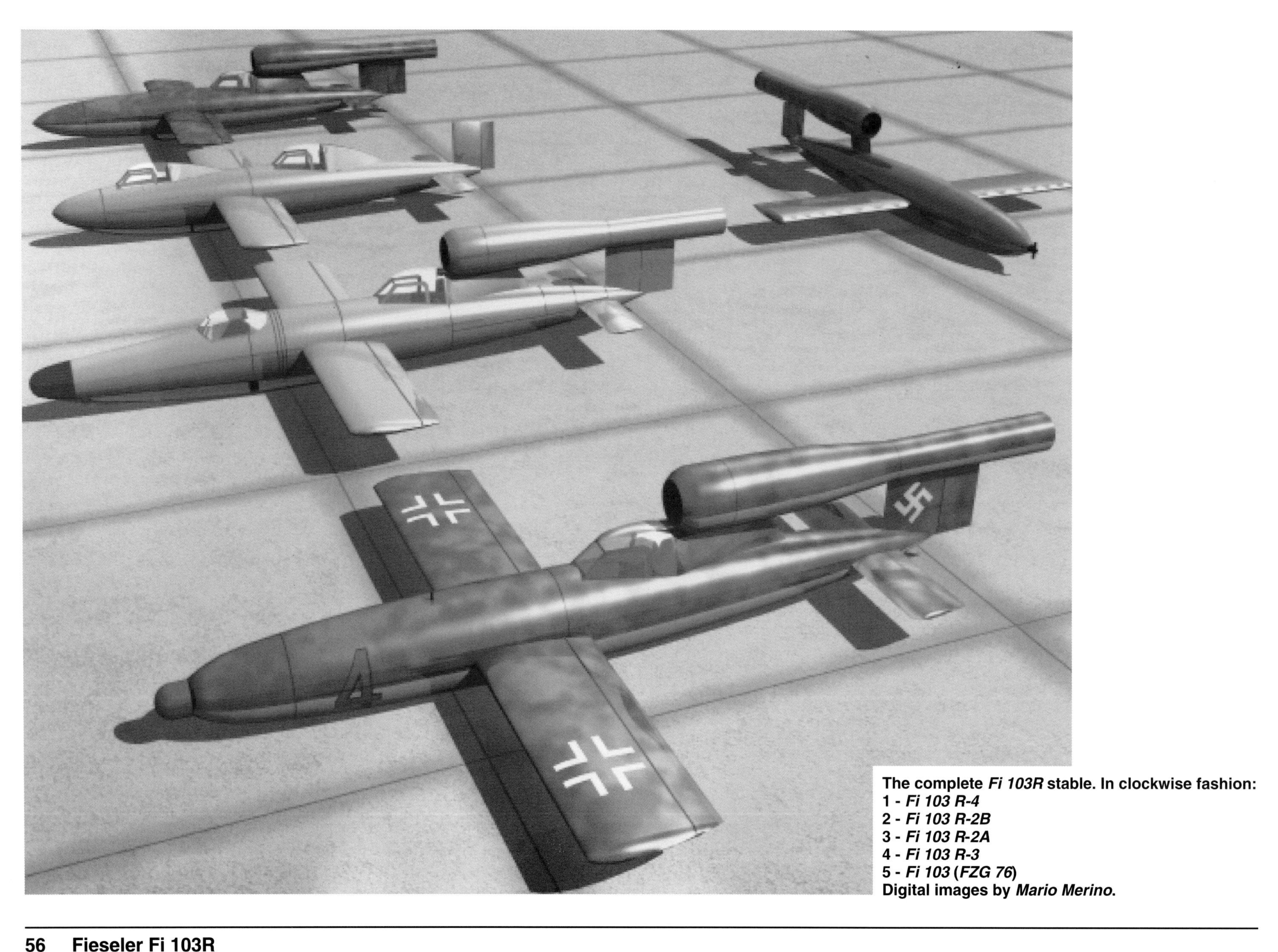

The complete *Fi 103R* stable. In clockwise fashion:
1 - *Fi 103 R-4*
2 - *Fi 103 R-2B*
3 - *Fi 103 R-2A*
4 - *Fi 103 R-3*
5 - *Fi 103* (*FZG 76*)
Digital images by *Mario Merino.*

The *Fi 103R-2A*, the unpowered tandem seat trainer as seen from its port side nose. Its overall length was shorter than the *Fi 103R-4* by several feet because the *Argus As 014* pulse jet engine was not needed in this initial trainer version. Digital image by *Mario Merino*.

A port side tail view of the unpowered tandem seat trainer *Fi 103R-2A*. The record is not entirely clear who built this primary trainer version of the *Fi 103R-4*. *Hanna Reitsch* implies that it was built by *DFS*-Ainring. Skorzeny implies that it was built by *Volkswagon*. Digital image by *Mario Merino*.

A head on view of the *Fi 103R-2A* unpowered tandem seat trainer. It is not known with certainty how many examples of this primary trainer were actually built. The American Army found several brand new two seat trainers post war with their plexiglass cockpit canopies purposely broken. Digital image by *Mario Merino*.

The *Fi 103R-2A* tandem seat trainer seen at high altitude. Its wooden landing skid was non retractable. It is not known with certainty if these primary trainers were conversions of the *FZG 76* flying bomb or if they were constructed out of wood. Digital image by *Mario Merino*.

A nice view of the tandem seat trainer sailplane *Fi 103R-2A* as it is about to pass over the camera. Construction of these dual-control trainers, indeed, the whole *Reichenberg* program was strictly secret. Digital image by *Mario Merino.*

A nose starboard side view of the *Fi 103R-2A* tandem seat trainer sailplane. The front cockpit occupied the space where the 1,874 pound *Amatol* warhead normally would have been located. Digital image by *Mario Merino.*

Fi 103R-2B. Argus As 014 pulse-jet powered tandem trainer featuring its starboard side. It is assumed that the dual control trainer would have been carried up to altitude by a *Heinkel He 111H* bomber and then released. This author is not aware of any documents outlining its flight performance. Digital image by *Mario Merino.*

A full starboard side view of the dual control *Fi 103R-2B* with its pulse-jet engine operating. This *Argus As 014* pulse jet equipped trainer carried sand bags to simulate the weight of the 1,874 pound *Amatol* warhead. Digital image by *Mario Merino.*

A port side view of an *Fi 103R-2B* powered dual control trainer. *Werner Baumbach's KG 200* was to have been the organization to operationally employ the *Fi 103R-4*. But all this changed in October 1944 when *Werner Baumbach* took charge of *KG 200* and changed his mind. Digital image by *Mario Merino*.

An overhead view of an *Fi 103R-2B* pulse-jet powered dual control trainer. It is thought that no flight training program was started before October 1944. It was at this time that *Oberstleutnant Werner Baumbach* took over as *Geschwader-Kommando* of *KG 200*. The entire suicide squadron was abandoned, and all flight instructors and volunteers reassigned. Digital image by *Mario Merino*.

The *Fi 103R-2B* powered tandem seat training machine as seen from its port side tail. Both *Skorzeny* and *Reitsch* were angry and felt betrayed when *Werner Baumbach* shut down all *Fi 103R* activities. But by October 1944 the Allied seaborne invasion at Normandy France was five months old. What targets of opportunity were left for the volunteer suicide pilots and their *Fi 103R-4s*? Digital image by *Mario Merino*.

An underside view of the tandem seat training aircraft *Fi 103R-2B* under full power. Once, said Hanna Reitsch, she suffered a hard landing during a solo flight in a dual seat trainer. The wooden landing skid was destroyed and the hull broke in two. She was not injured. Digital image by *Mario Merino*.

The *Fi 103R-2B* tandem seat trainer seen from its starboard side at altitude and full power. *Hanna Reitsch* realized that a pilot of average ability would have great difficulty in safely landing the dual control trainers and that most would not survive on their own. Digital image by *Mario Merino.*

The *Fi 103R-3* single seat powered trainer as seen from its port side. Notice that its wooden landing skid is only about one half as long as found on the *Fi 103R-2B* tandem seat trainer version. Water ballast was still used in place of the 1,874 pound *Amatol 39* warhead. Digital image by *Mario Merino.*

A nice overhead view of the *Fi 103R-3* single seat training aircraft featuring its port side. Test flights may have been made with this training machine, because photographs exist which purportedly show the machine in flight. Digital image by *Mario Merino*.

A port side rear view of the single seat trainer *Fi 103R-3*. This version, like the dual control trainer, had landing flaps...which the *Fi 103 R-4* did not have, because it would have been a one way flight. Digital image by *Mario Merino*.

A nose on view of the *Fi 103R-3* single seat trainer aircraft featuring its basic design simplicity. No *Amatol* warhead was fitted, and water ballast was used to simulate the weight of the warhead. Digital image by *Mario Merino*.

A full underside view of the single seat pilot trainer *Fi 103R-3*. Pretty straightforward and simple design. *Hanna Reitsch* said piloting these machines was at all times an extremely difficult and dangerous operation. Once she had piloted one in a dive up to 530 miles per hour. Digital image by *Mario Merino*.

The real thing...the *Fi 103R-4* in flight. It did not have landing flaps or a wooden landing skid, so its first flight would be its last. Its nose-mounted 1,874 pound *Amatol 39* warhead has been painted red. Digital image by *Mario Merino.*

A full top view of the single seat pilot trainer *Fi 103R-3.* Digital image by *Mario Merino.*

The *Fi 103R-4* flying out front and beneath. The entire trailing edge of the wing was an aileron. Digital image by *Mario Merino*.

The *Fi 103R-4* in a dive. The machine carried sufficient fuel for 32 minutes of powered operation, at which time it would have traveled approximately 204 miles. Digital image by *Mario Merino*.

An overhead view of the German "*kamikaze*" *Fi 103R-4*. Maximum forward speed was an estimated 497 miles per hour at 8,000 feet. Digital image by *Mario Merino*.

One of the few well persevered *Fi 103R-4*s on display in the world is located at the *Technical Museum* at Delft, Holland. It has been faithfully restored and painted. Photographed by *Ed Straten*.

A close-up view of the *Fi 103R-4*. Photographed by *Ed Straten*.

A close-up view of the *Fi 103R-4's* port side cockpit. Notice the cramped quarters. The pilot's head rests on the plywood seat back, inches beneath the *Argus As 014's* air intake. The cockpit hood was manufactured as a single unit and was hinged to open to starboard. Photographed by *Ed Straten.*

Looking down at the *Fi 103R-4's* port side cockpit. The top of the cockpit canopy was a single piece of plexiglass. Inside the cockpit can be seen the thick armored glass windscreen. To the far lower left can be seen the wing root and aileron located on the wing's trailing edge. Photographed by *Ed Straten.*

A view of the *Fi 103R-4's* upper fuselage area forward the armored glass windscreen. The thickness of this armored glass can be judged when looking into the windscreen (window). Photographed by *Ed Straten*.

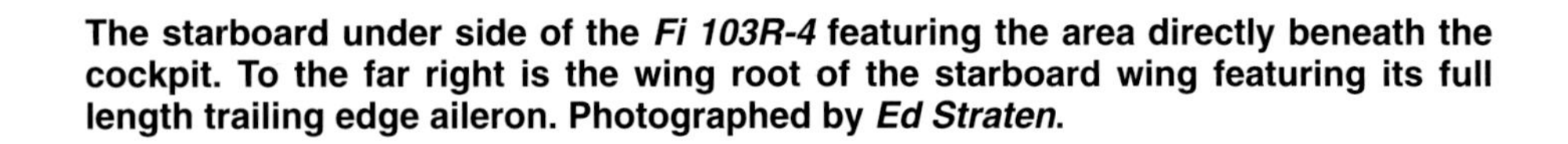

The starboard under side of the *Fi 103R-4* featuring the area directly beneath the cockpit. To the far right is the wing root of the starboard wing featuring its full length trailing edge aileron. Photographed by *Ed Straten*.

The *Fi 103R-4's* port side fuselage featuring the streamlined fairing between the Argus As 014 pulse jet engine and the machine's aft tapered fuselage. The steel metal band and cable helps suspend the machine to the museum's roof trestles. Photographed by *Ed Straten*.

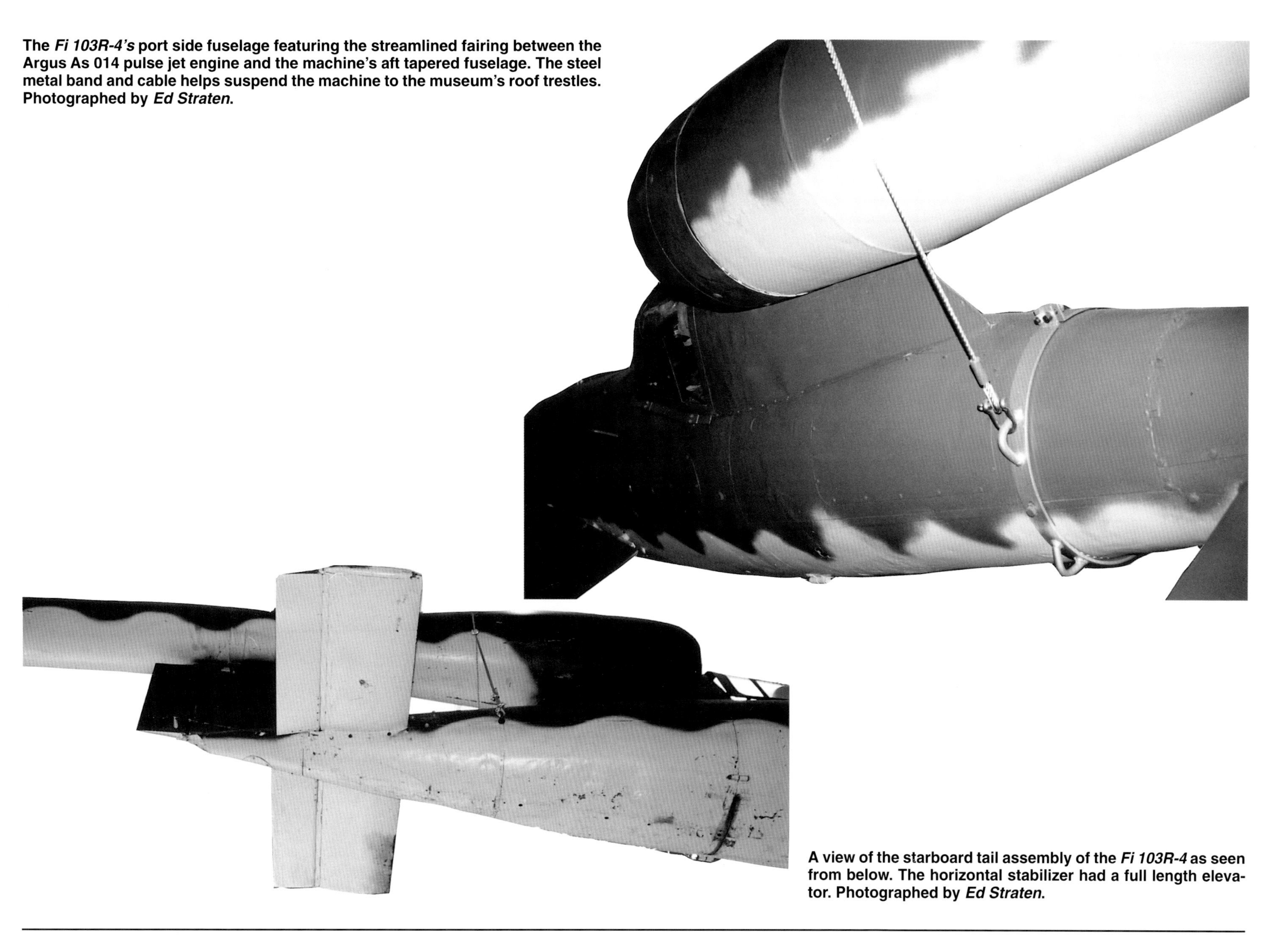

A view of the starboard tail assembly of the *Fi 103R-4* as seen from below. The horizontal stabilizer had a full length elevator. Photographed by *Ed Straten*.

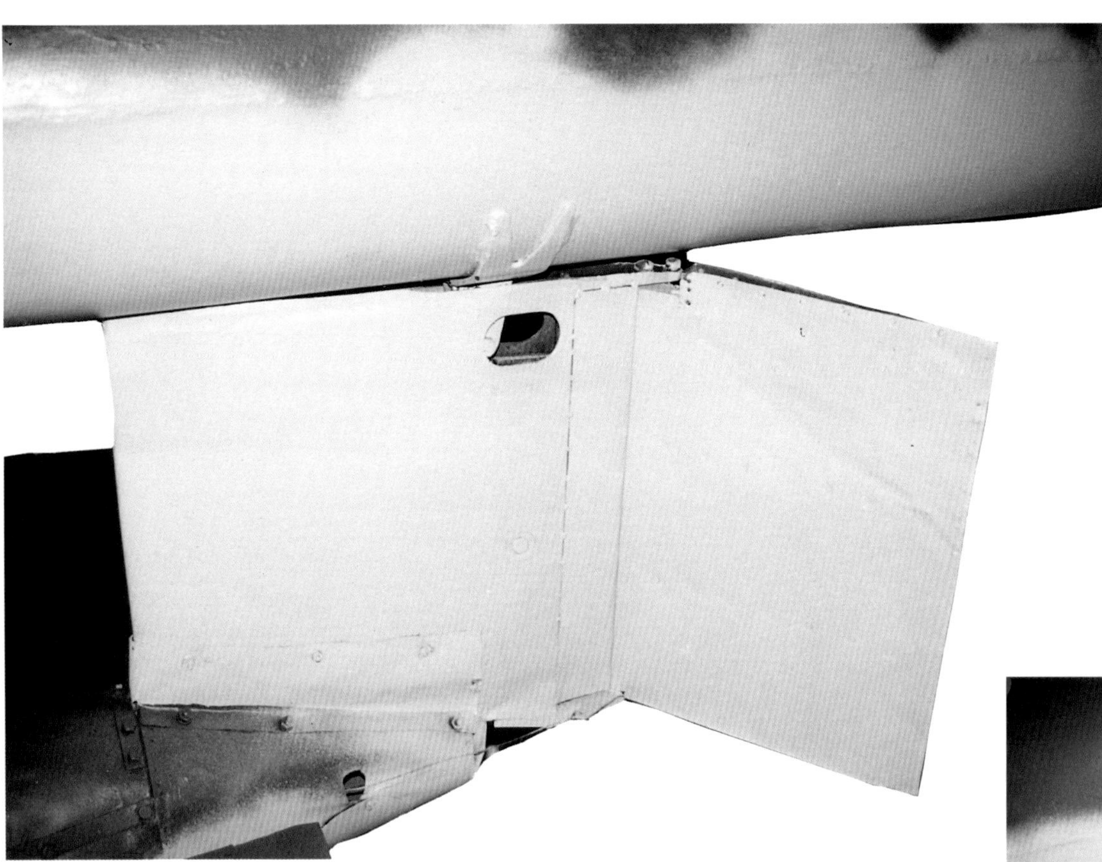

The movable rudder as seen on the port side *Fi 103R-4*. The rudder was articulated by a small diameter wire cable attached to the articulating rudder lever located near the lower piano-type rudder hinge. The cable then passed along on the outside of the tapered fuselage through a small diameter round hole and on the pilot's foot-operated pedals. Photographed by *Ed Straten*.

The rudder on this *Fi 103R-4* has been articulated hard to port..the full range of its movement to port. Photographed by *Ed Straten*.

A close-up the port side rudder fin, piano-type hinged rudder, rudder articulating metal lever, rudder articulating cable, tail cone, and upper surface of the horizontal stabilizer's elevator as found on the *Fi 103R-4*. Photographed by *Ed Straten*.

A close-up of the port side rudder fin, rudder articulating cable running from inside the tail cone to the rudder articulating lever attached to the piano-type hinged rudder as found on the *Fi 103R-4*. Photographed by *Ed Straten*.

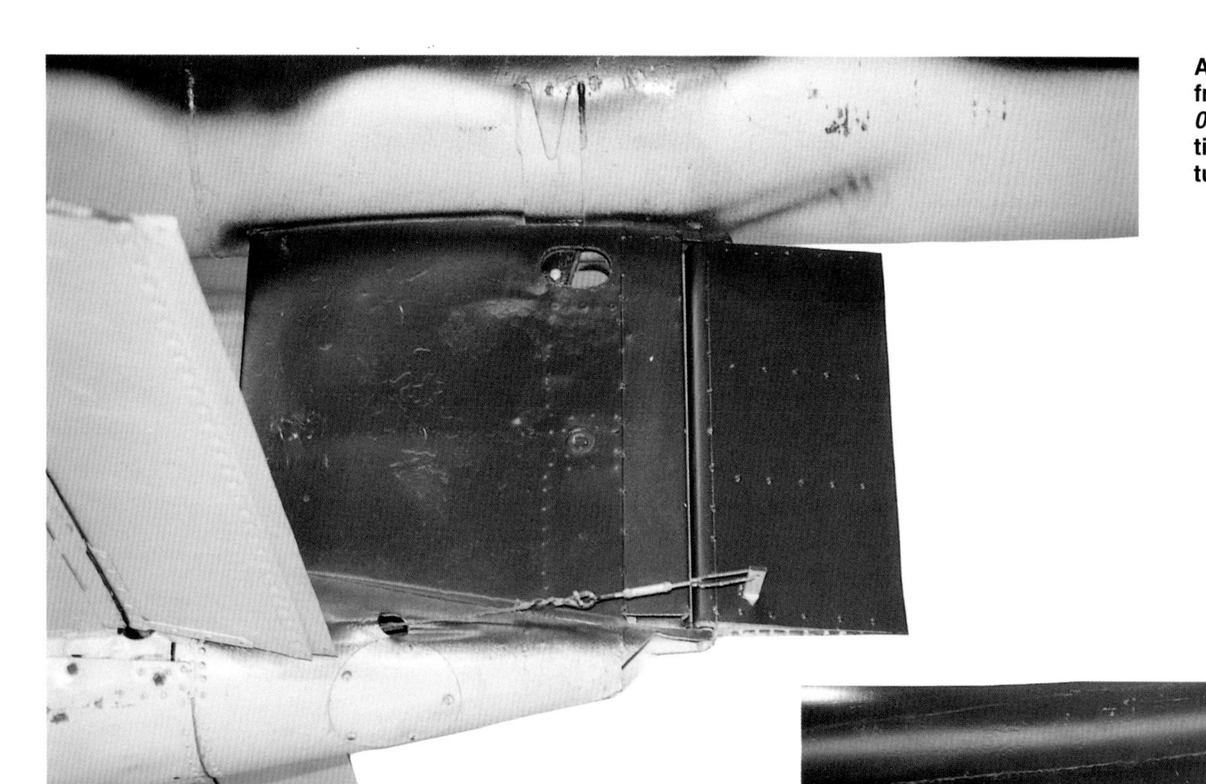

A view of the port side rudder assembly on an *Fi 103R-4* as seen from beneath. Featured at the top of the photo is the *Argus As 014* jet pipe, rudder fin, piano-type hinged rudder with zero articulation, rudder articulating lever, rudder articulating cable with turnbuckle, tail cone, and elevators. Photographed by *Ed Straten*.

The full port side tail assembly of an *Fi 103R-4*. Seen at the top of the photograph is the *Argus As 014* jet pipe, rudder fin, piano-type hinged rudder, horizontal stabilizer with elevator, aft fuselage and fuselage tail cone. Photographed by *Ed Straten*.

The *Argus As 014's* jet pipe and tail assembly of an *Fi 103R-4* as seen from its port side. The piano-type hinged rudder is articulated full to port. Photographed by *Ed Straten.*

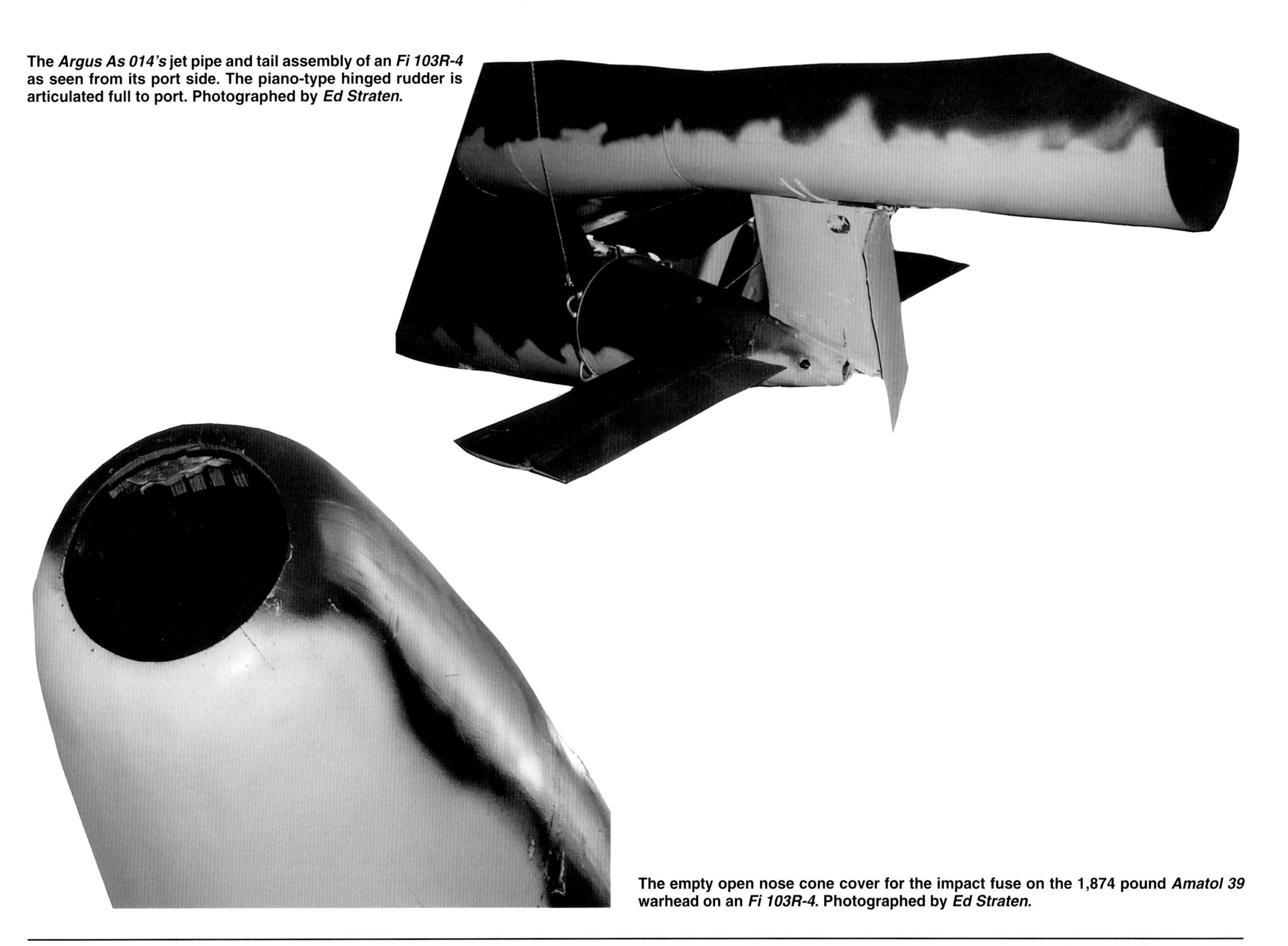

The empty open nose cone cover for the impact fuse on the 1,874 pound *Amatol 39* warhead on an *Fi 103R-4*. Photographed by *Ed Straten.*

The single spark plug of the *Argus As 014* pulse jet engine, which is found on the upper surface of the combustion chamber section of the jet engine. Photographed by *Ed Straten.*

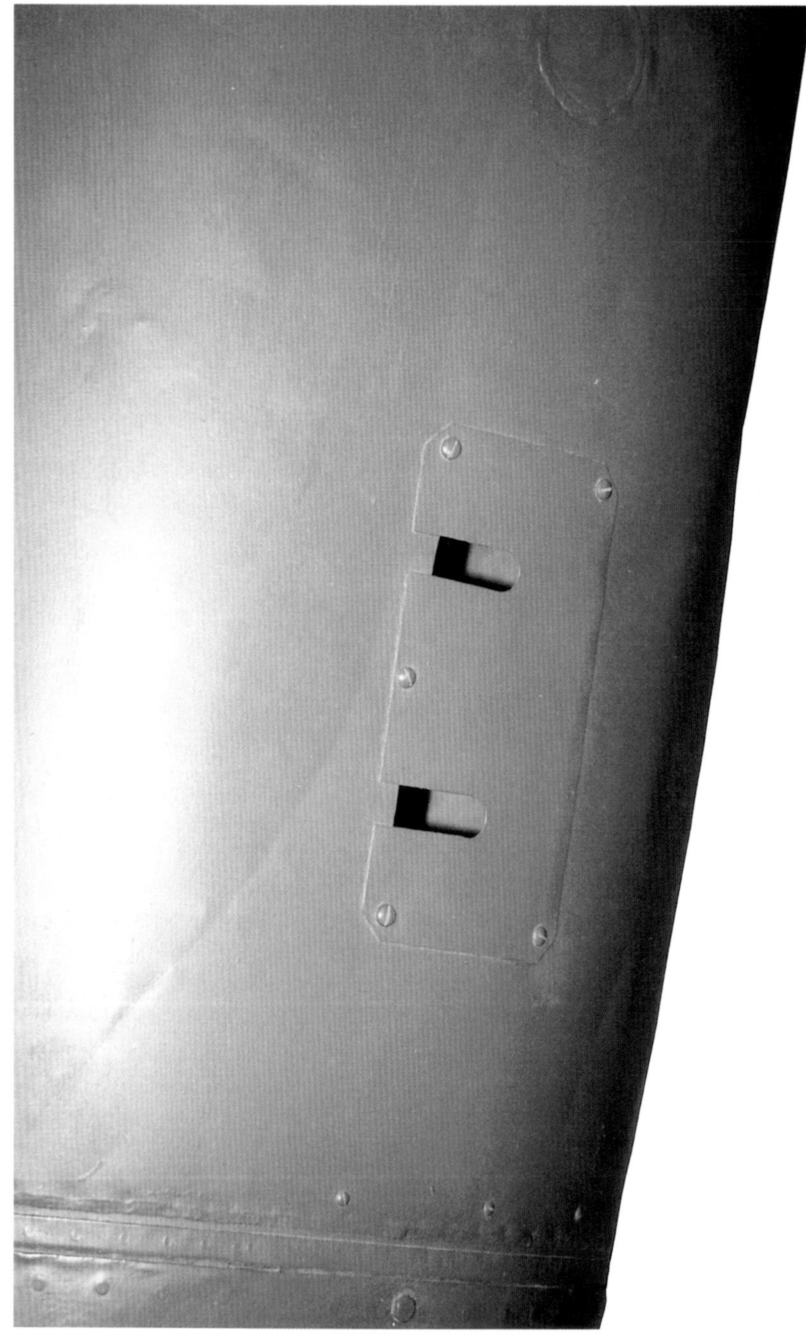

The forward upper fuselage located fuse sockets for arming the *Amatol 39* warhead on the robot *Fi 103A-1* prior to catapult launching. They were not used nor covered over during its conversion into the manned *Fi 103R-4*. Photographed by *Ed Straten*

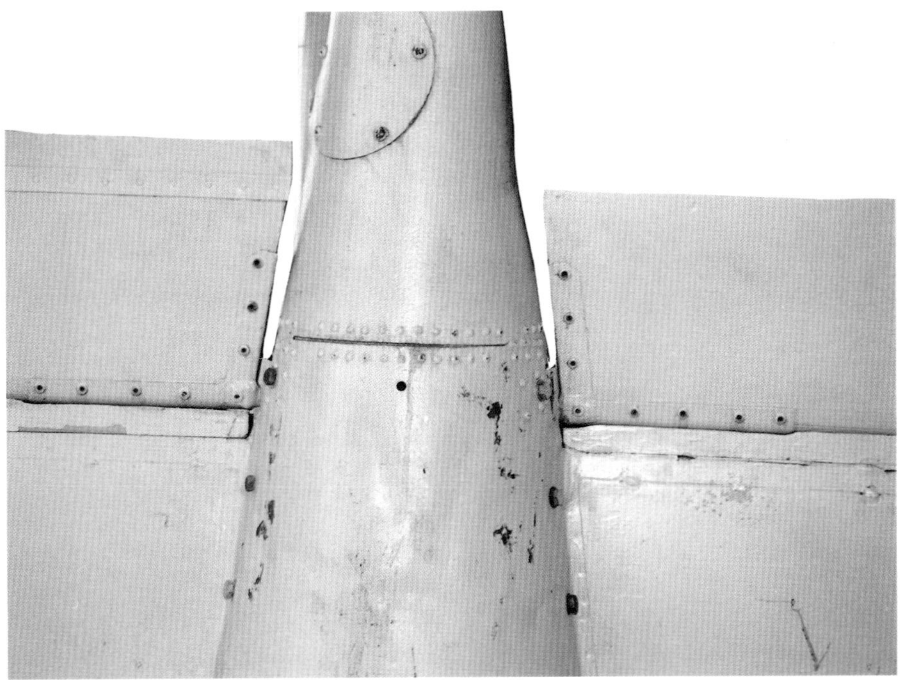

The underside surface of an *Fi 103R-4's* tail assembly as seen from beneath and featuring the aft tapered fuselage, port elevator (left side of photograph), and starboard elevator. Photographed by *Ed Straten*.

The port side wing root, wing trailing edge with hinged aileron, inboard aileron hinge, aileron articulation lever, and fuselage-mounted Pitot tube. Photographed by *Ed Straten*.

0-50

Opposite, Left: Upper fuselage of an *Fi 103R-4* featuring the bolted together sections containing the fuel tank and the section containing the 1,874 pound *Amatol 39* warhead. The small diameter metal tube contained the warhead arming impact switch wire. The round cap is the fuel filler port. Aft the fuel filler port is the lifting lug port. At the lower bottom of the photo are the port and starboard wing roots. The silver metal band and cable are used to suspend the *Fi 103R-4* from the museum ceiling. Photographed by *Ed Straten.*

Opposite, Right: Upper starboard side leading edge wing root of an *Fi 103R-4*. Photographed by *Ed Straten.*

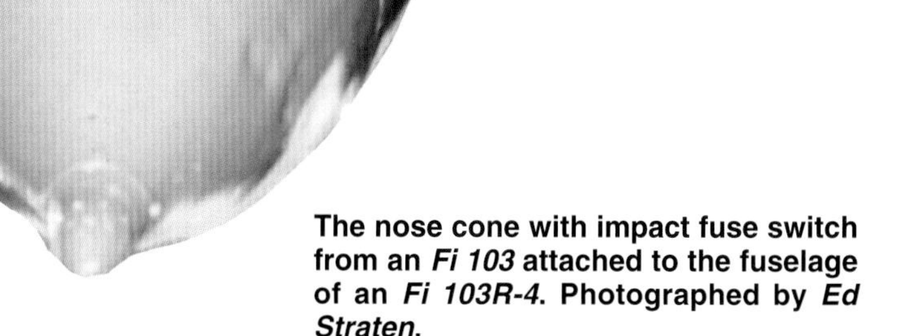

The nose cone with impact fuse switch from an *Fi 103* attached to the fuselage of an *Fi 103R-4*. Photographed by *Ed Straten.*

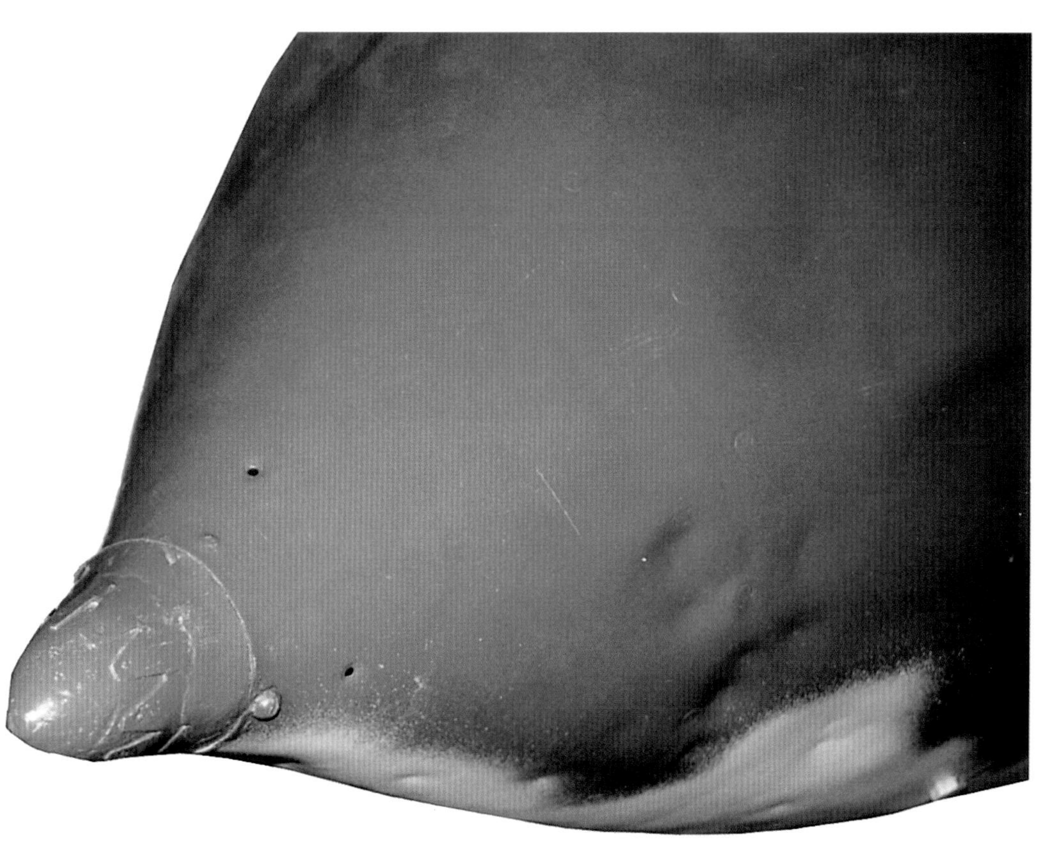

A close-up view of the nose cone with impact fuse switch from an *Fi 103*. Photographed by *Ed Straten.*

An underside view of an *Fi 103R-4* featuring its starboard side wing panel, trailing edge aileron, aileron articulating lever in the wing root, inboard aileron hinge, and the trailing edge locking pin device. Photographed by *Ed Straten*.

An underside view of an *Fi 103R-4* featuring the fuselage and starboard side wing, wing root, and inboard aileron hinge. The *Fi 103* and *Fi 103R-4* utilized a tubular spar. The wing panels were attached by sliding each panel over the tubular spar. The small device seen on the wing's undersurface mid-way up the wing root is the trailing edge's locking pin device. Photographed by *Ed Straten*.

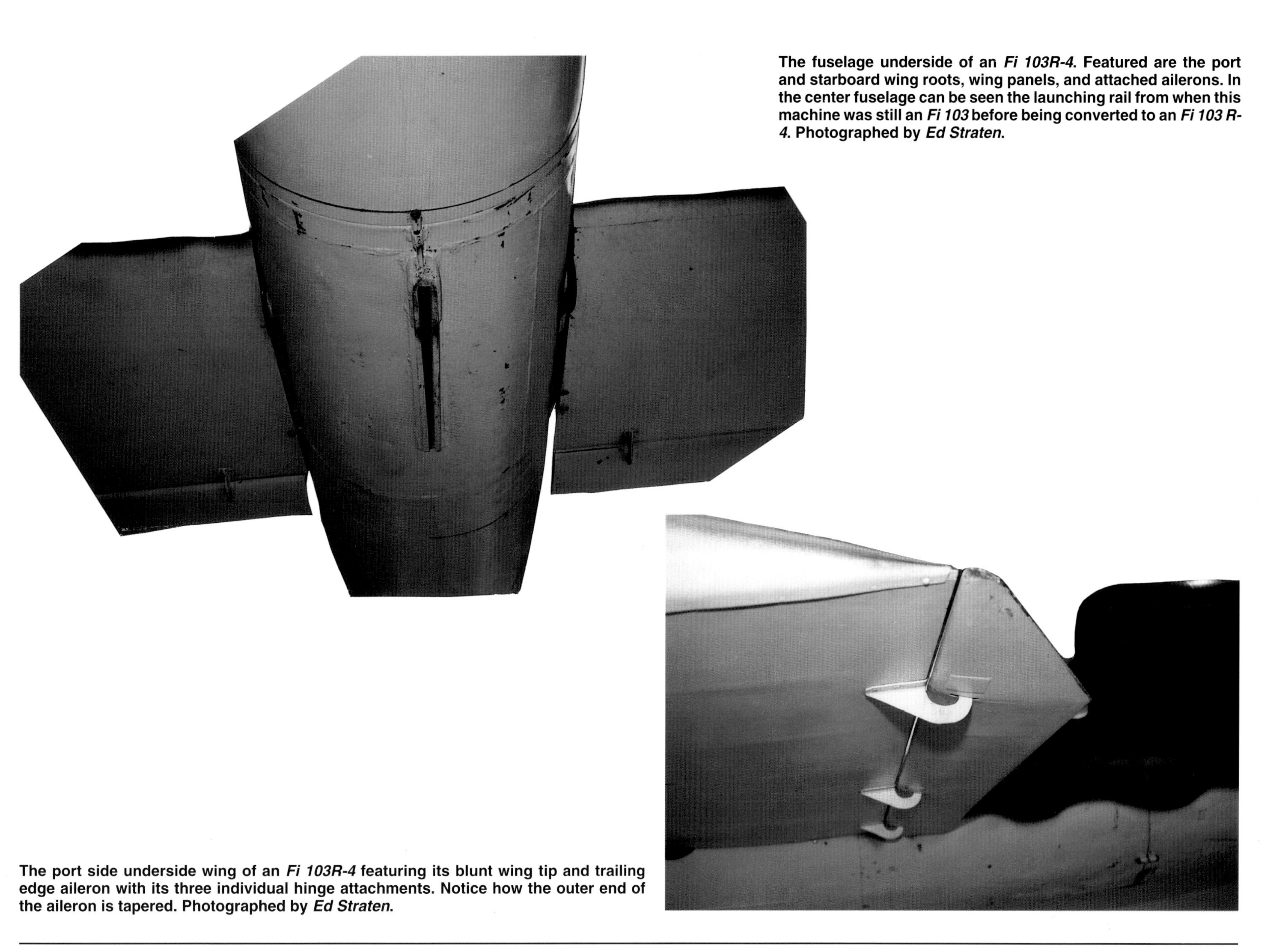

The fuselage underside of an *Fi 103R-4*. Featured are the port and starboard wing roots, wing panels, and attached ailerons. In the center fuselage can be seen the launching rail from when this machine was still an *Fi 103* before being converted to an *Fi 103 R-4*. Photographed by *Ed Straten*.

The port side underside wing of an *Fi 103R-4* featuring its blunt wing tip and trailing edge aileron with its three individual hinge attachments. Notice how the outer end of the aileron is tapered. Photographed by *Ed Straten*.

A port side view of an *Fi 103R-4* featuring its blunt wing tip. The round hole in the wing tip is an access port to secure the wing panel to its round metal pipe-like spar. Photographed by *Ed Straten*.

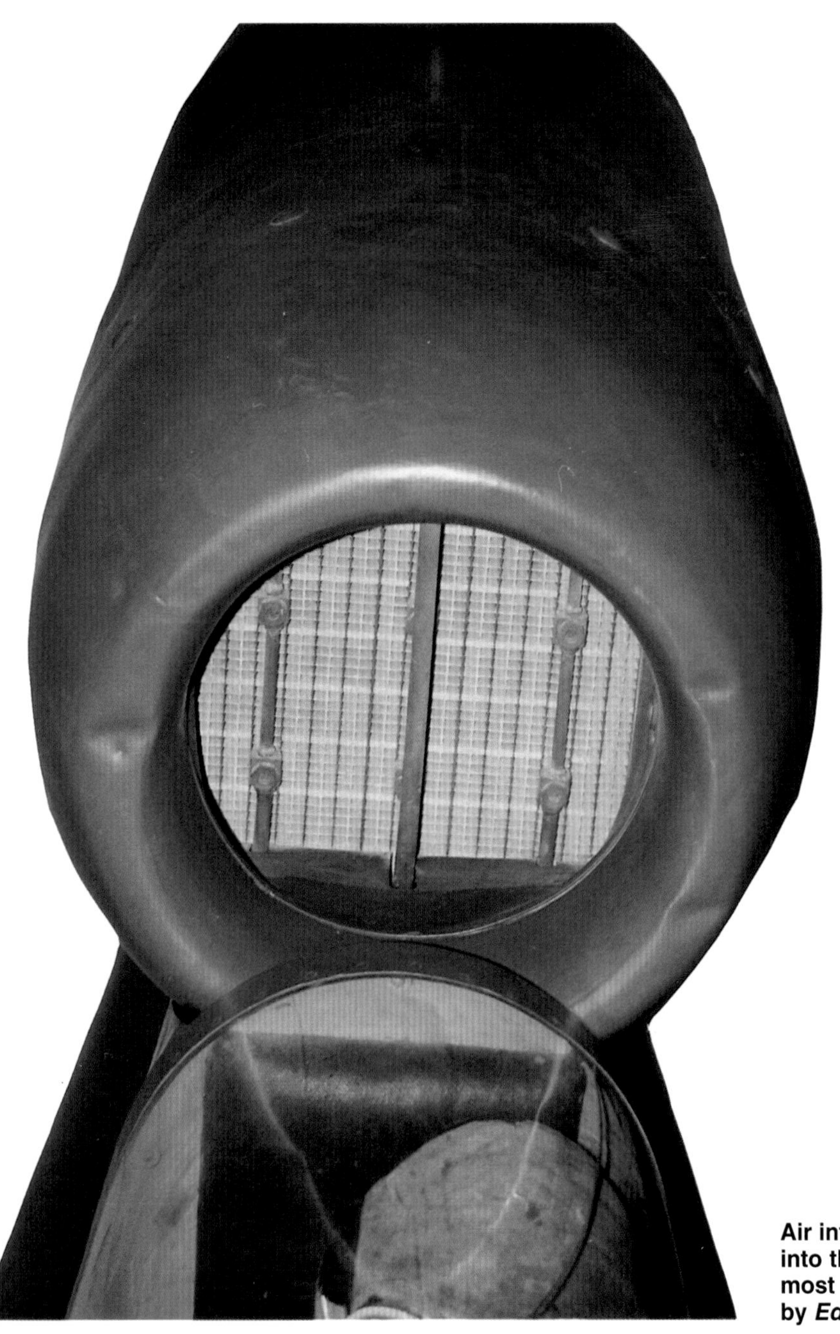

Air intake of the *Argus As 014* pulse jet on the *Fi 103R-4*. The cockpit canopy's top extended up into the air intake opening. This is one reason why *Skorzeny*, *Reitsch*, and others believed that most pilots would not be able to jump out of the machine prior to hitting its target. Photographed by *Ed Straten*.

A port side view of an *Fi 103R-4* featuring its wing leading edge. The original *Fi 103 V-1* had a cable wire cutter inside the leading edge. Since the *Fi 103R-4* is a conversion, the cable wire cutter is included behind the leading edge. Photographed by *Ed Straten*.

A port side view of the *Fi 103R-4* appearing as if in flight. Photographed by *Ed Straten*.

A nice close-up of an *Fi 103R-4's* port side cockpit cabin with the volunteer pilot looking right at the camera. Photographed by *Ed Straten.*

A port side view of the aft fuselage and tail assembly featuring its horizontal stabilizer on an *Fi 103R-4*. The silver colored metal band with wire cable is to suspend the machine from a roof trestle at the museum. Photographed by *Ed Straten.*

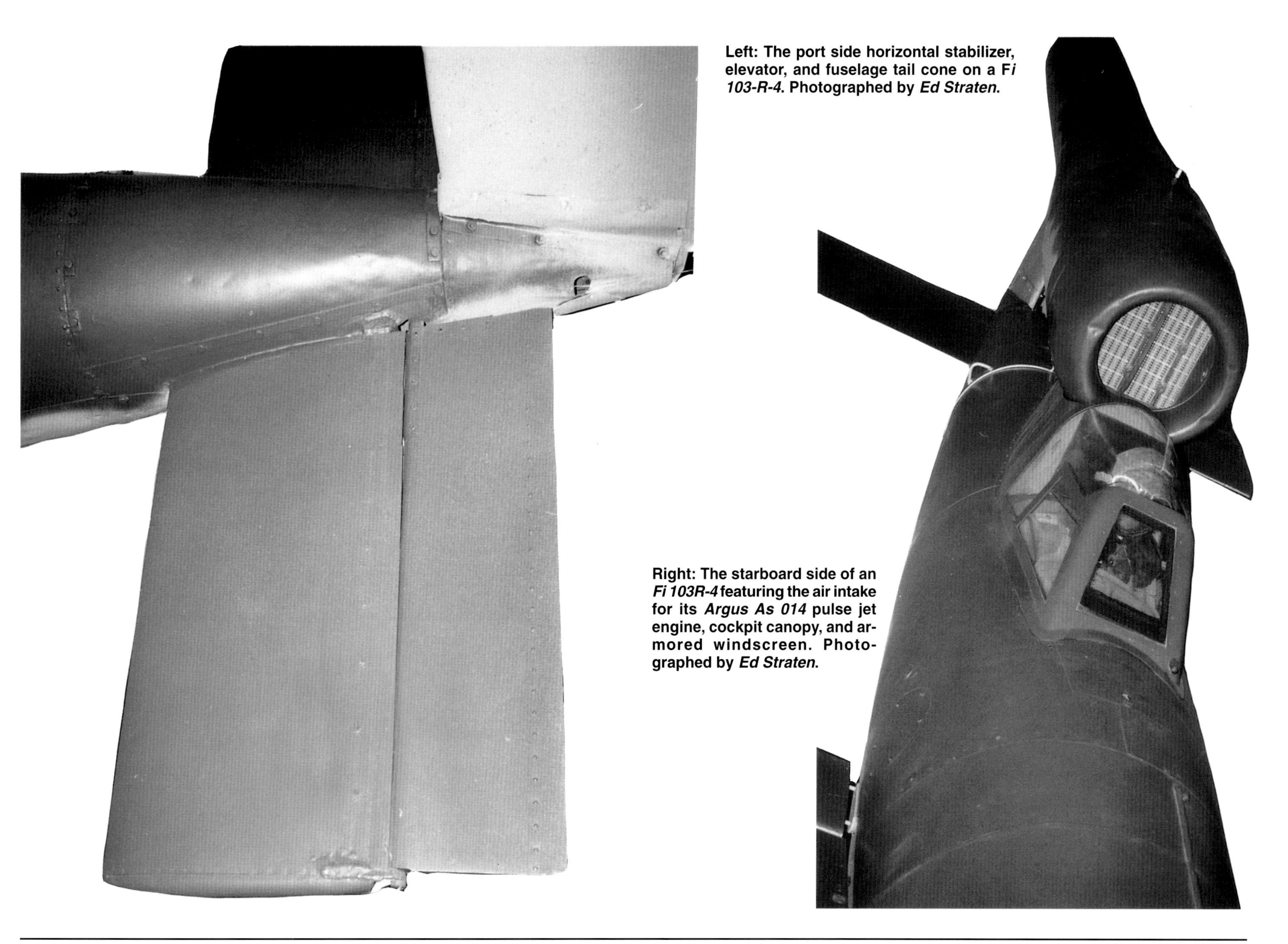

Left: The port side horizontal stabilizer, elevator, and fuselage tail cone on a *Fi 103-R-4*. Photographed by *Ed Straten*.

Right: The starboard side of an *Fi 103R-4* featuring the air intake for its *Argus As 014* pulse jet engine, cockpit canopy, and armored windscreen. Photographed by *Ed Straten*.

A port side close up view of an *Fi 103R-4's* leading edge wing, wing root, cockpit, and air intake of its *Argus As 014*. Photographed by *Ed Straten*.

Port side view of the warhead section drawn and bolted on to the fuel tank section of a F*i 103R-4*. On the under side of the fuel tank section can be seen its launching rail for use on a catapult launcher when the machine was first constructed as a *Fi 103 V-1*. Photographed by *Ed Straten*.

A poor quality photo of the *Fi 103 V-1's* under fuselage launch rail as seen on an *Fi 103R-4*. Photographed by *Ed Straten*.

An underside view of the other port wing on an *Fi 103R-4* featuring its tapered aileron and outboard hinge and hinge bracket. Photographed by *Ed Straten*.

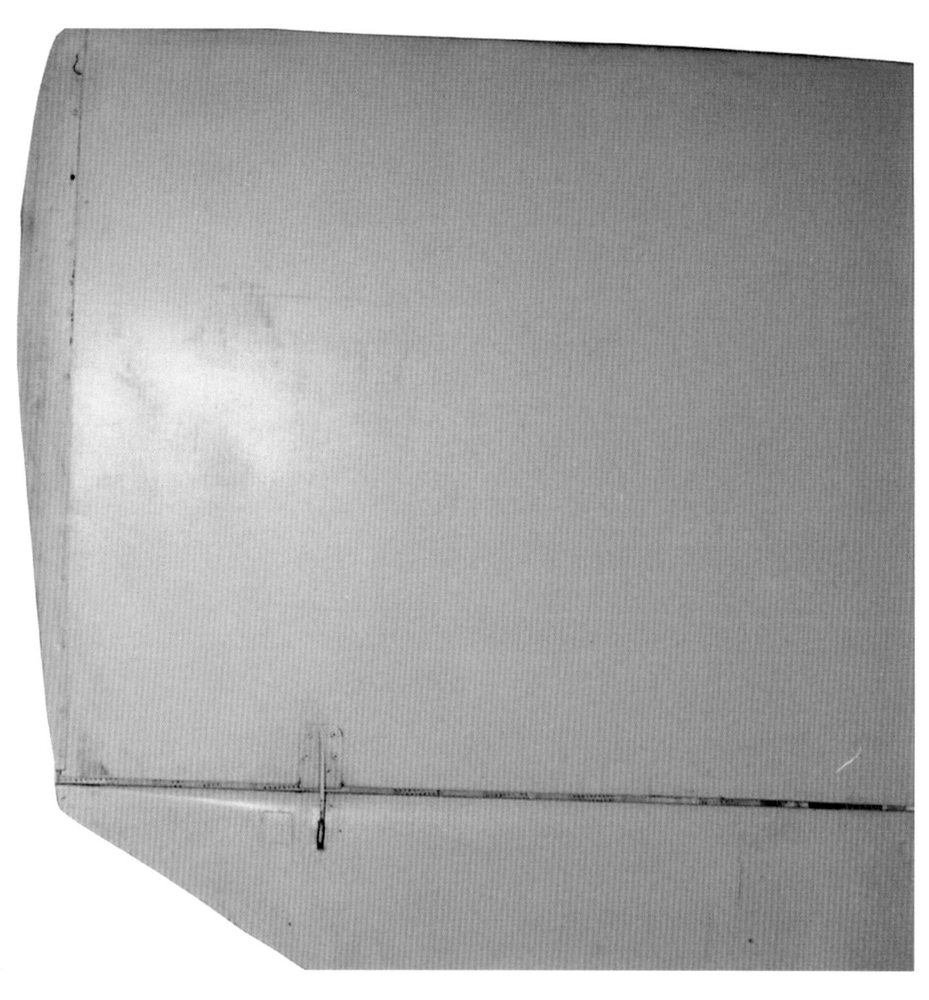

The port wing tip, tapered aileron, and hinge of an *Fi 103R-4* as viewed from beneath. Photographed by *Ed Straten*.

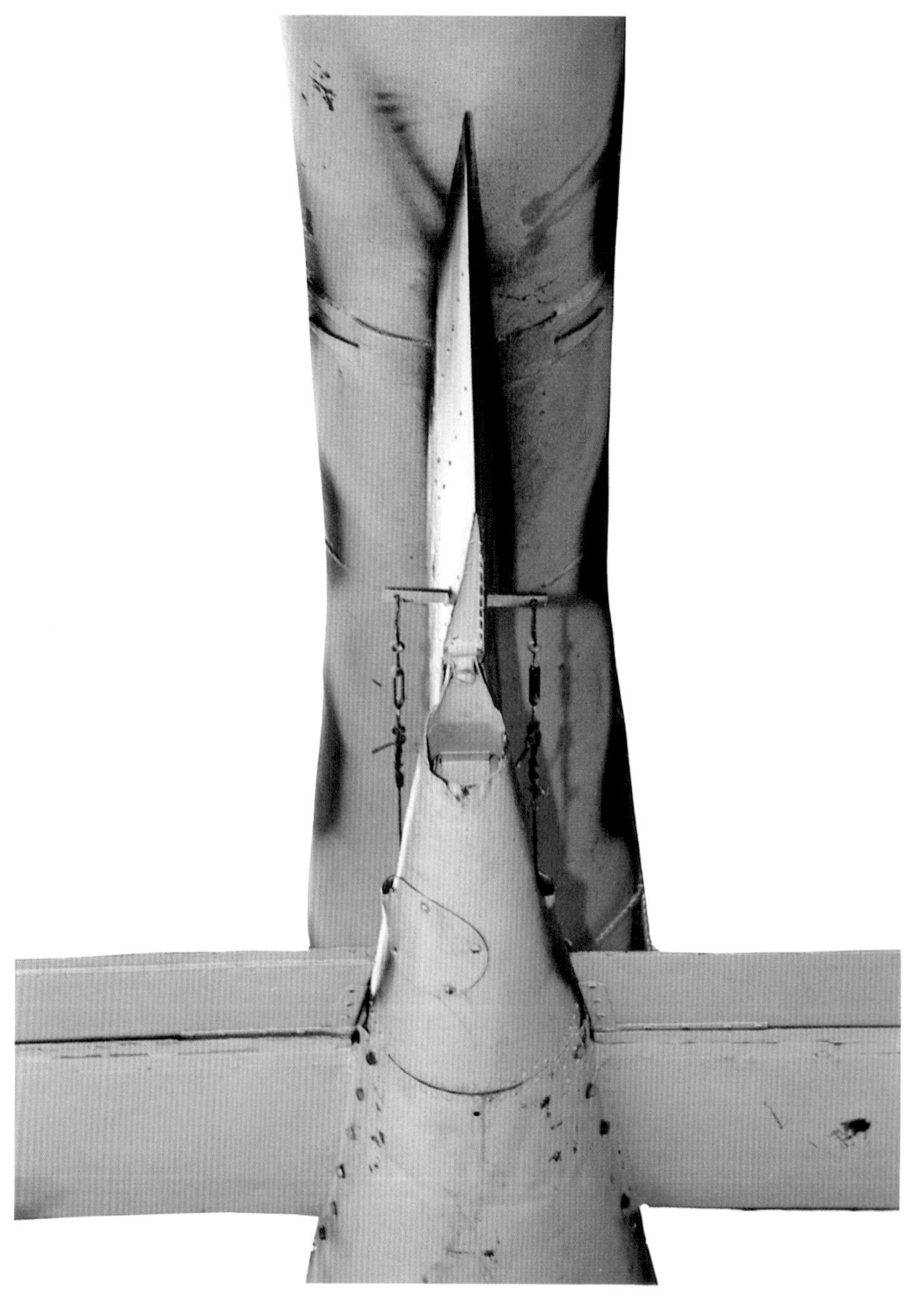

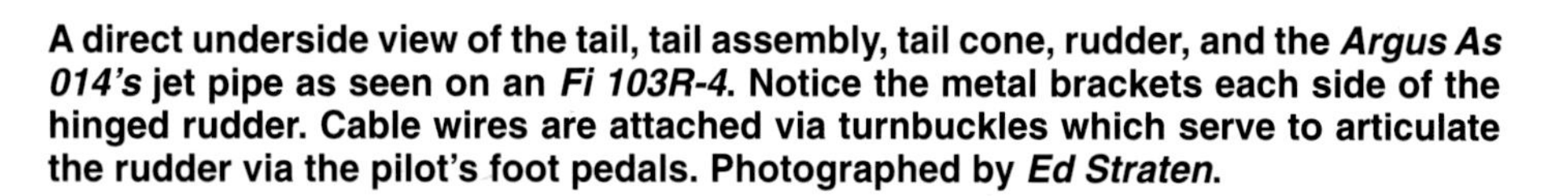

A direct underside view of the tail, tail assembly, tail cone, rudder, and the *Argus As 014's* jet pipe as seen on an *Fi 103R-4*. Notice the metal brackets each side of the hinged rudder. Cable wires are attached via turnbuckles which serve to articulate the rudder via the pilot's foot pedals. Photographed by *Ed Straten*.

A view of an *Fi 103R-4's* rear fuselage, however, its hinged rudder has been articulated full to port. The metal articulating brackets with its cable attachment are evident in the photo. Photographed by *Ed Straten*.

A freshly converted *Fi 103 V-1* to an *Fi 103R-4* found by six American troops. The metal pipe along the lower starboard side is its tabular spar. Its *Amatrol 39* warhead is covered over by a wooden nose cone for shipping. Prior to flight it would be replaced with a metal nose cone.

An American Army "Technical Sargent" (left) peers into the *Fi 103R-4's* cockpit while a colleague removes the one-piece cockpit canopy. The seat back inside the cockpit was constructed out of plywood.

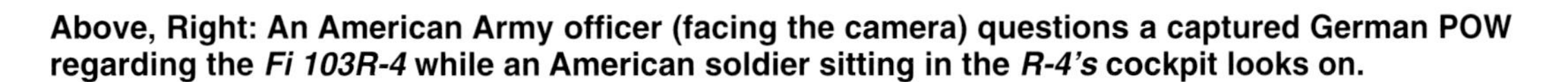

Above, Right: An American Army officer (facing the camera) questions a captured German POW regarding the *Fi 103R-4* while an American soldier sitting in the *R-4's* cockpit looks on.

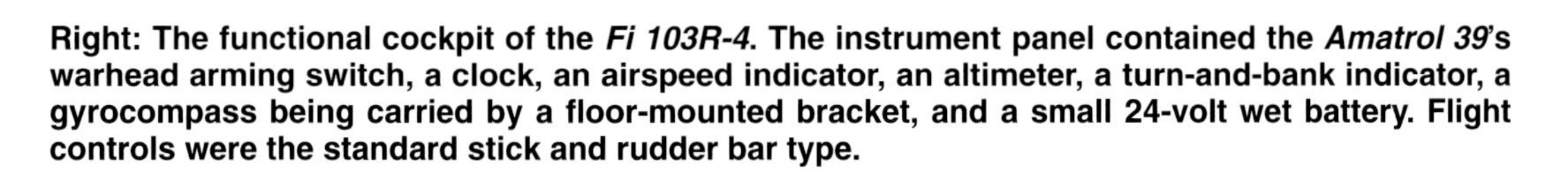

Right: The functional cockpit of the *Fi 103R-4*. The instrument panel contained the *Amatrol 39*'s warhead arming switch, a clock, an airspeed indicator, an altimeter, a turn-and-bank indicator, a gyrocompass being carried by a floor-mounted bracket, and a small 24-volt wet battery. Flight controls were the standard stick and rudder bar type.

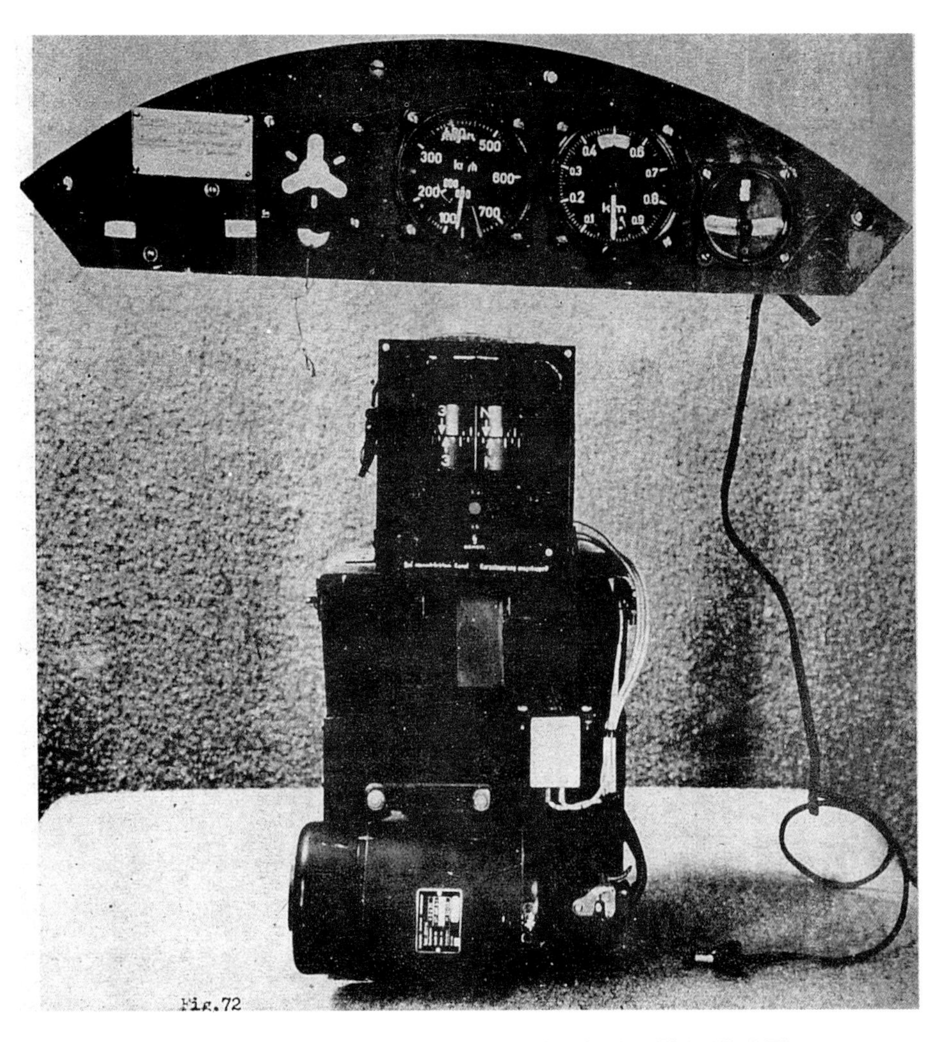

The complete instrument panel prior to installation in the *Fi 103R-4*. The gyrocompass fit between the pilot's legs.

A poor quality photo of an American Army service man sitting in the cockpit of an *Fi 103R-4*. Notice how the cockpit canopy opens to starboard.

Converted *V-1s* into *Fi 103R-4s* in storage as found by the American Army. Their *Amatrol 39* warheads are covered with a protective plywood nose cone.

Above, Right: American Army *Colonel Anderson* looks over new *Fi 103R-4s* found in storage at a railhead.

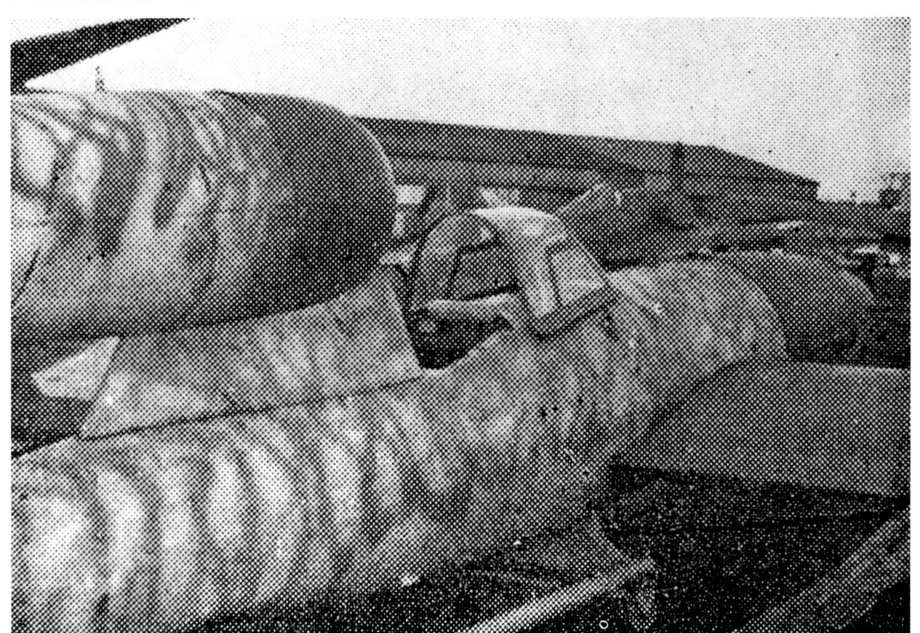

Right: A poor quality photo of an *Fi 103R-4* in England post war. The machine's starboard side is featured.

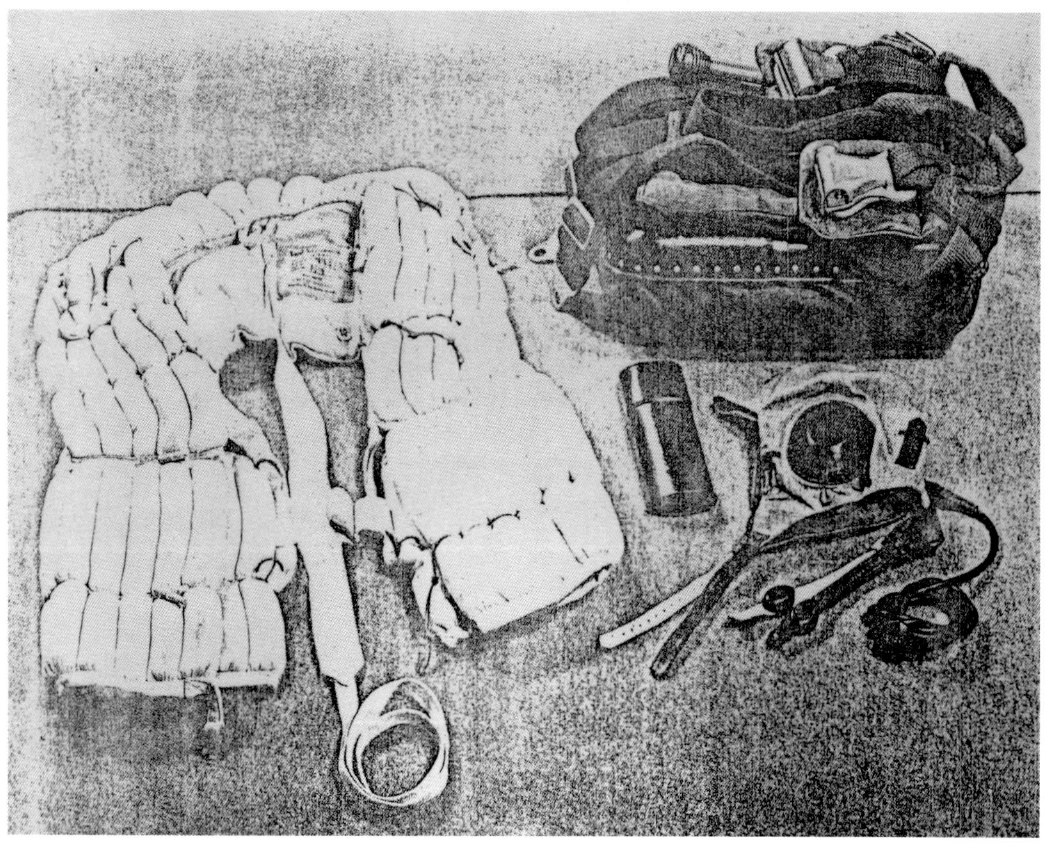

A poor quality photo of an American soldier dressed up in an *Fi 103R-4* pilot's gear, such as life jacket, leather flight cap, goggles, earphones, and parachute.

Above, Left: A fully assembled *Fi 103R-4* set up on a tarp for a photo session post war. Its starboard side is shown in the photograph.

Left: The pilot of an *Fi 103R-4* would have worn these items on his so-called suicide flight. Left to right: life jacket, parachute, flare, leather flight cap with earphones, and goggles.

An *Fi 103R-4* on static public display during the German Aircraft Exhibition at RAF-Farnborough, England, November 1945. It is believed that his was the only *Fi 103R-4* brought to England post war.

A poor quality starboard side view of the RAF-Farnborough's *Fi 103R-4*. November 1945.

An American soldier tries out the fit and comfort in the cockpit of an *Fi 103R-4* as seen from its nose port side. A plywood plug has been inserted in the nose where the impact fuse for the *Amatrol 39* warhead would normally have been located.

A nose on view of the *Fi 103R-4* with an American soldier sitting in its cockpit. The railroad seen to the right of the photo is how these *suicide* machines would have been transported to *KG 200's* airstrip to be loaded on *Heinkel He 111Hs* for air launching. Approximately 125 were found in storage at the rail head ready to be shipped.

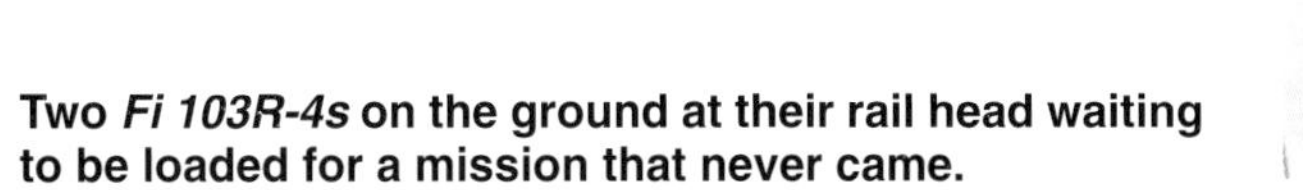

Two *Fi 103R-4s* on the ground at their rail head waiting to be loaded for a mission that never came.

A poor quality photo of a new *Fi 103R-4* as seen from its starboard side ready for shipping. Its tubular spar is attached to the machine. Its wings would also be loaded next to the *R-4* prior to shipping.

A poor quality photo of the starboard side of an *Fi 103R-4*. Its location is probably Wright-Patterson Air Force Base, Dayton, Ohio.

An Americanized *V-1* known as the *VB-2* seen on outdoor public display at the Air Force Exhibit Unit, Wright-Patterson Air Force Base, Dayton, Ohio. On its port side is painted "Ist Guided Missile Group."

In addition to the *Argus As 014* powered *Messerschmitt Me 328* and the *Fieseler Fi 103R*, several other German aircraft designers were considering the pulse jet as a power source. This included *Heinkel* and his proposed twin *Argus As 014* powered *He 164A-10* and the single *Argus As 014* pulse jet powered *He 162A-11*. Neither machine was flown with the *Argus As 014* by war's end. Scale model and photographed by *Steve Malikoff*.

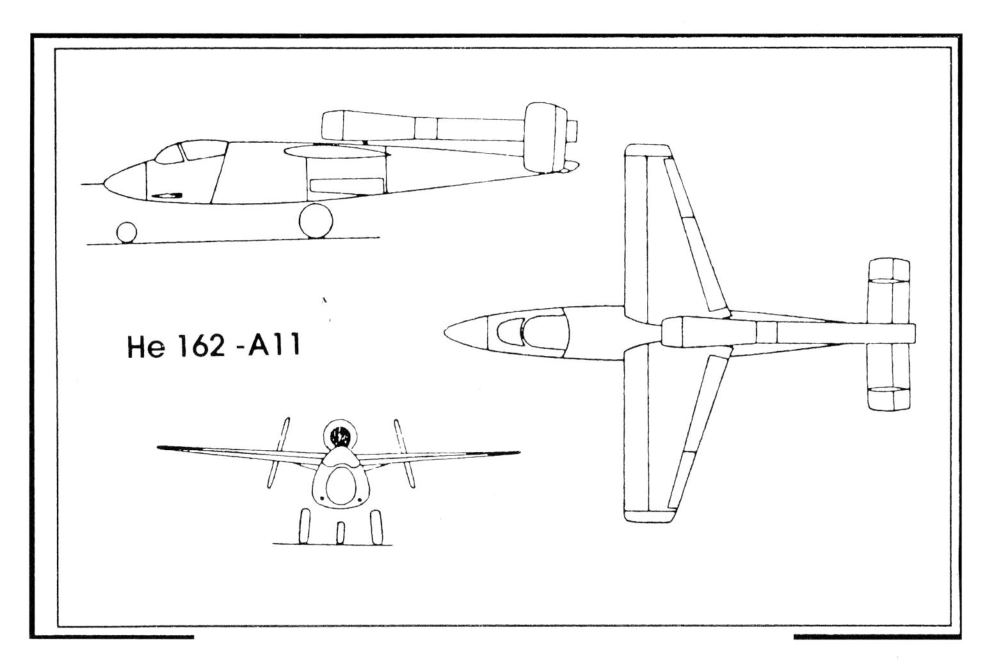

A pen and ink 3-view drawing of the proposed single *Argus As 014* pulse jet powered *Heinkel He 162A-10.*

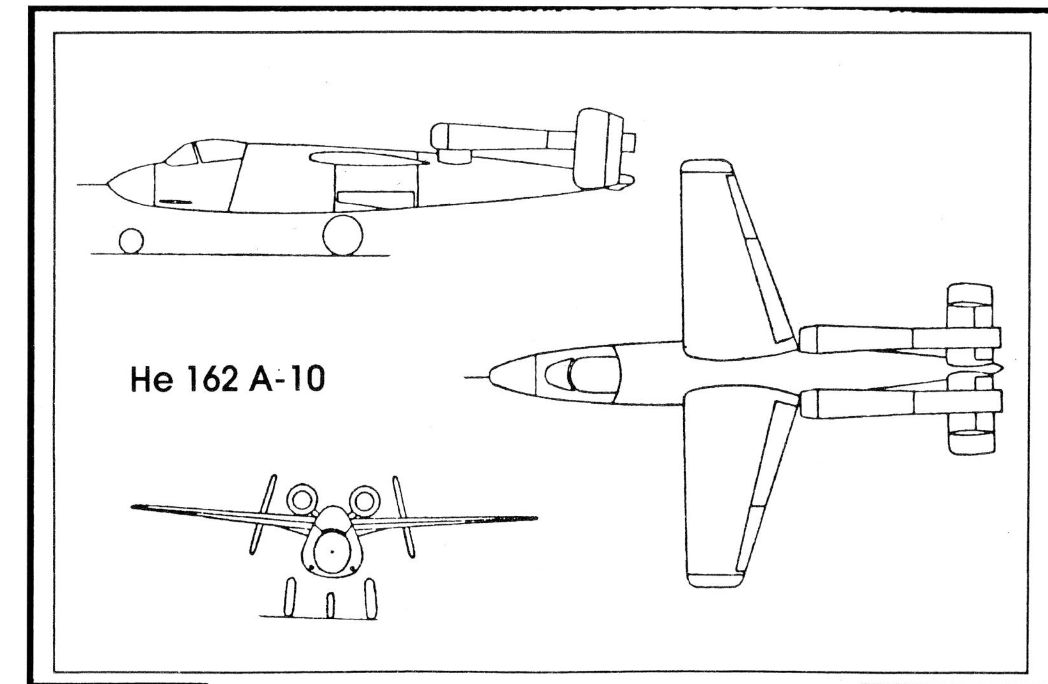

A pen and ink 3-view drawing of the proposed twin *Argus As 014* pulse jet powered *Heinkel He 162A-11.*

A pen and ink 3-view drawing of the proposed single *Argus As 014* pulse jet powered *Blohm und Voss Bv 213* miniature fighter. The intent was to construct a small fighting machine in the most economical manner in terms of material and man-hours.

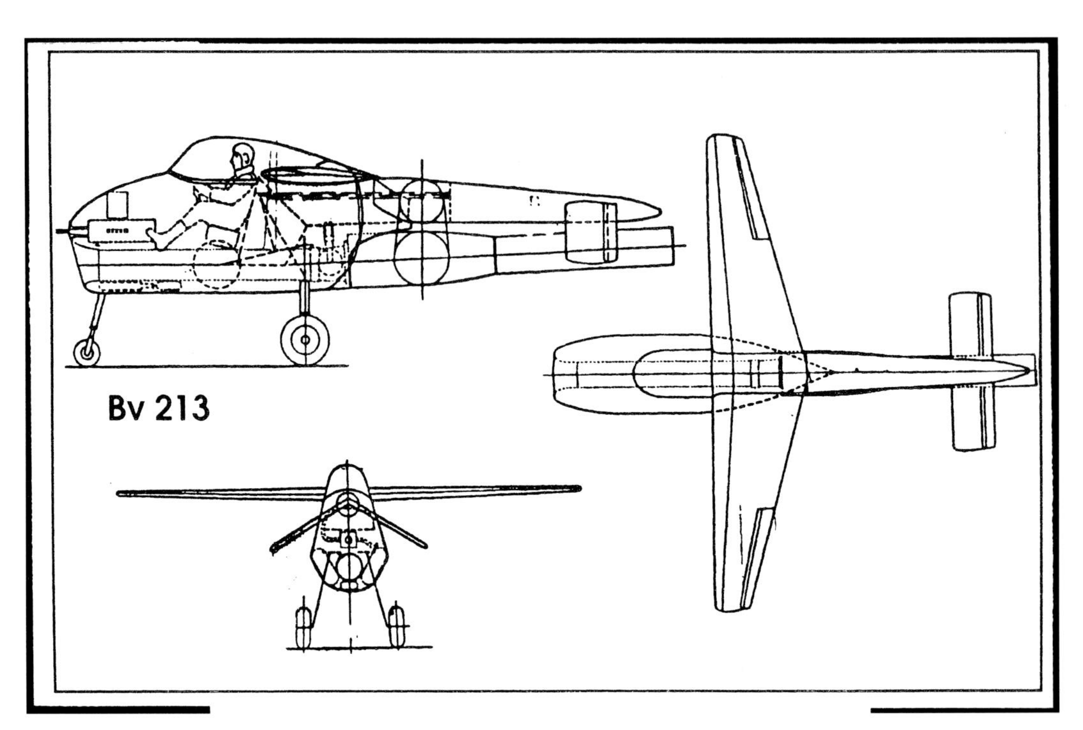

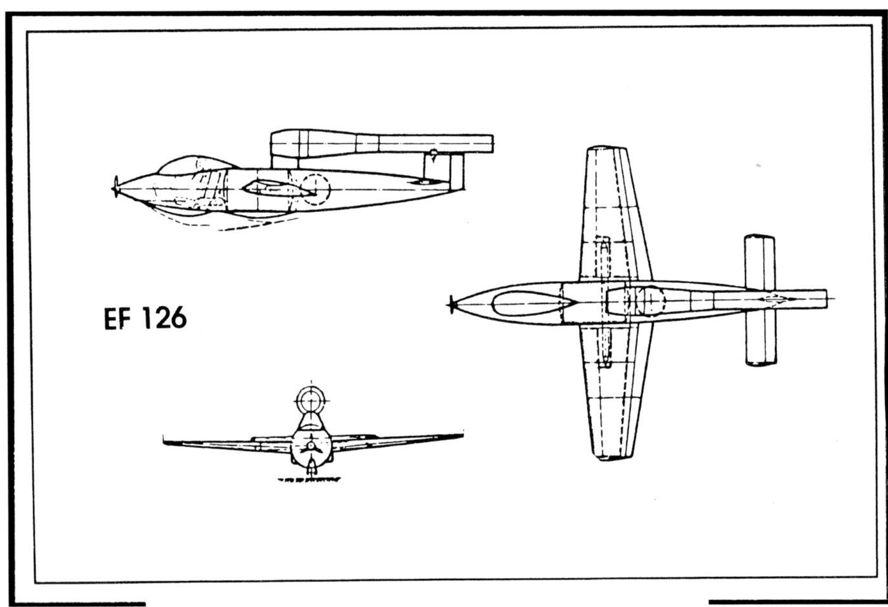

A pen and ink 3-view drawing of the proposed single *Argus As 014* pulse jet powered *Junkers Flugzeugbau Ju EF 126* "*Elli*" ground attack aircraft.

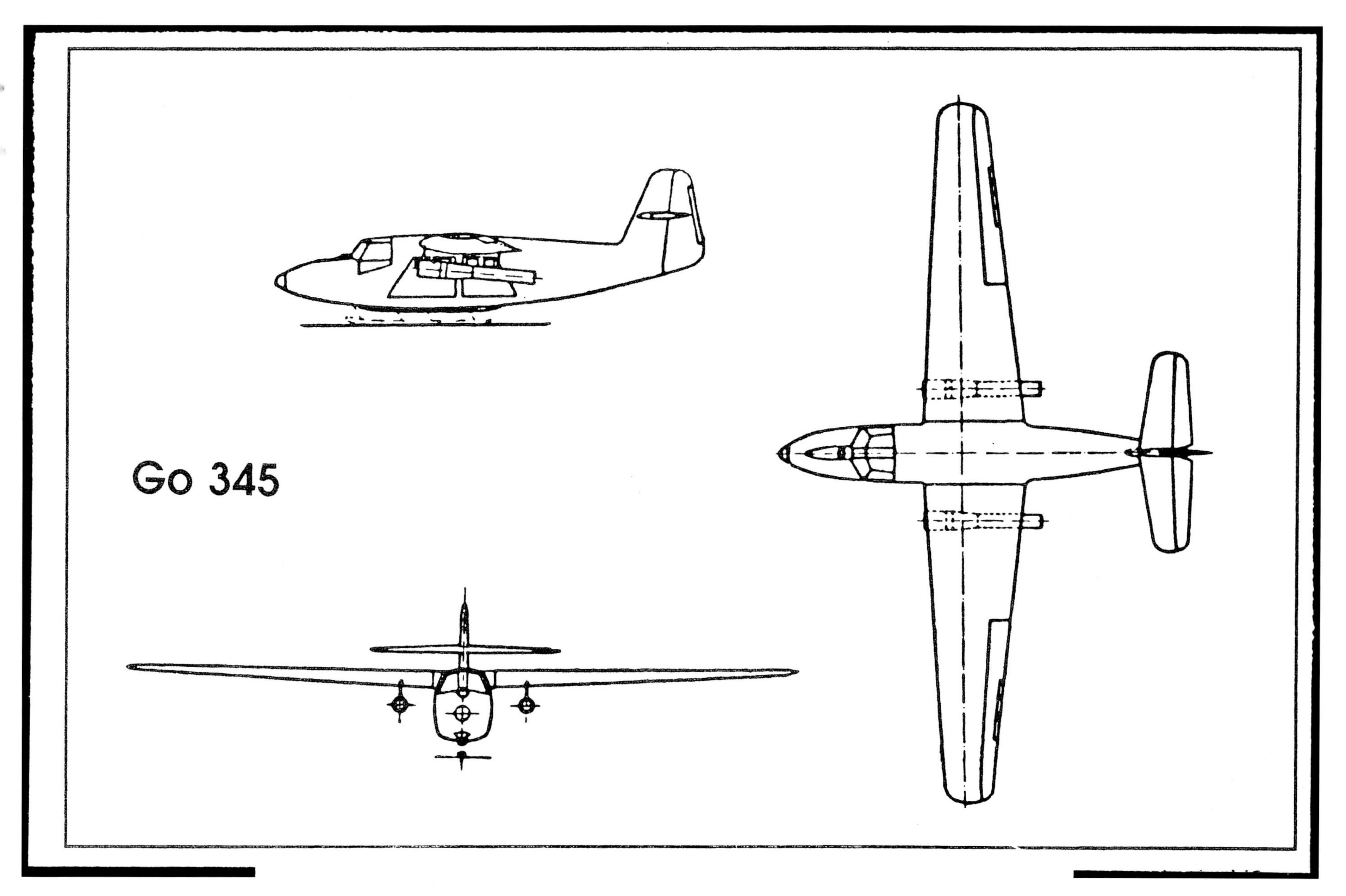

A pen and ink 3-view drawing of the proposed twin *Argus As 014* pulse jet powered *Gothaer Waggonfabric* (*Gotha*) *Go 345* troop carrier assault aircraft.

Index

A

Allied Normandy Invasion - 6-8, 15-16, 62
Amatol 39 warhead - 4, 59-60, 63, 65-66, 75-76, 79, 90-91, 93, 95
Anderson, Colonel - 93
Arado Ar 96B - 8
Argus As 014 - 3-6, 12-14, 36-37, 44, 46, 54-55, 57, 60, 69, 71, 74-76, 82, 85-86, 88, 98-101

B

Balka - 16
Battle for Britain -34
Baumbach, Oberstleutnant Werner - 15, 61-62
Blohm und Voss Bv P.213 - 100
Braun, Eva - 9
Benzinger, Dr.-Med. Theo - 13

C

CK+[illegible]AN - 27

D

D-Day - 3
Deutsche Forschungsinstitut für Segelflug-Ainring (DFS) - 3, 7-8, 12, 14-15, 34, 36, 52, 57
DFS 230 - 8
Dittmar, Heinrich "Heini"- 8
Doodlebug - 3
Donnerberg factory - 15
Dornier Do 17 - 8
Dornier Do 217 - 12, 37
Deutschlandhalle-Berlin - 31

E

El Duce - 40
Elli - 100
Erprobungsstelle-Rechlin - 12

F

Fieseler, Gerhard - 12, 20-21
Fieseler, Luthur - 12
Fieseler Fi 103 A-1 V1 - 3, 5-6, 8-10, 12-13, 20-21, 23-26, 28-29, 43-44, 48-49, 53, 76, 79-81, 83, 86-87, 90
Fieseler Fi 103 V1 VB-2 - 98
Fieseler Fi 103R - 3, 5, 8, 12, 14-15, 56, 62, 98
Fieseler Fi 103R-2A - 13, 53-54, 56-59
Fieseler Fi 103R-2B - 54, 56, 60-63
Fieseler Fi 103R-3 - 12, 53, 55-56, 63-66
Fieseler Fi 103R-4 - 5, 6, 9-13, 16, 30, 45-47, 50-51, 53, 55, 57, 61-62, 64, 66-77, 79-88, 90-97
Flying Club-Berlin - 13
Flying Is My Life - 9
Focke-Achgelis-Delmenhorst Fa 61 - 7, 30-31
Focke, Professor - 7
Focke-Wulf Flugzeugbau - 9
Focke-Wulf Fw 190 - 14-15, 38
Fort Ebel Emael - 8
Fritz X - 30
FZG 76 Type 1 - 3-4, 13, 15, 21-24, 41-43, 46-47, 51, 56, 58
FZG 76 Type 2 - 22, 27, 33

G

Gerhard Fieseler Werke - 12
Goebbels, Joseph - 3, 15-16, 33
Göring, Reichsmarshall Hermann - 7-8, 14-16, 33, 35
Georgii, Professor Walter - 7, 14-15, 34, 52
Gothaer Waggonfabrik Go 345 - 101
Gran Sasso Mountains - 4

H

Habicht - 34
Hegl, Oberst - 14
Heinkel, Professor Ernst - 9
Heinkel AG - 10, 98
Heinkel He 111H - 10-13,15, 27-28, 48-49, 60, 96
Heinkel He 162A-10 - 98-99
Heinkel He 162A-11 - 98-99
Heinkel He 177 - 34
Himmler, SS Reichsführer Heinrich - 9, 14
Hitler, Adolf - 3, 8-9, 14, 16, 34-35, 39
Hilter Bunker - 33
Hirth, Wolf - 6
Hollow Charge Warhead - 39
HMS Warspite - 30
Horten brothers - 34
Horten Ho 7 - 20
HWK steam catapult - 23-24, 26

I

Iron Cross - 8, 35
I-10 - 18
I-18 - 20

J

Jacobs, Hans - 7
Junkers Flugzeugbau Ju EF 126 - 100

K

Kampfgeschwader 33 - 4
Kampfgeschwader 200 - 14-15, 61, 96
Kampfgeschwader 5/KG 200 - 14
Kamikaze - 9, 11-12, 14-15, 37, 38, 47, 68
Kensche, Heinz - 13
Korton, General - 14
Kracht, Felix - 15

L

Lange, Leutnant Heinrich - 13-14
Lashendon Air Warfar Museum - 51
Lindbergh, Charles - 7
Lorin, Rene - 3
Luftwaffe - 3, 7-9, 11, 14, 32, 34, 47, 54

M

Madelung - 7
Marconnet - 3
Matthews, Trevor - 51
Medical Aeronautics-Rechlin - 13
Messerschmitt Me 163B - 8, 10
Messerschmitt Me 328 - 8, 12, 14-15, 36-37, 98
Merino, Mario - 3
Milch, Feldmarshall Erhard - 8-11, 15, 40-42, 47, 49, 51
Mistel - 15, 38-39
Mitsubishi G4M2E - 17-18

Mussolini, Benito - 3-4, 39, 40
MXY-8 "Oka 11" - 16-20

N
Nazis - 14, 16, 27
Nazism - 9
New York Times - 32

O
Oka 11 - 16-20, 23
Oka 22 - 20
Operation Reichenberg - 10-11
Operation Suicide - 13

P
Porsche, Professor - 9

R
RAF-Farnborough - 95
Radl - 48
Rechin - 7
Reichenberg, Code name - 12, 15, 59
Reichenberg-1 - 3, 15, 52
Reichenberg-2 - 3, 15, 52
Reichenberg-3 - 3, 15, 52
Reichenberg-4 - 3, 5, 15, 52
Reichsluftfahrtministerium (RLM) - 3, 7, 9, 11, 14
Reitsch, Hanna - 3, 5-6, 8-11, 13-16, 21, 30-36, 38, 41, 50-52, 57, 62-63, 65, 82
Rheinst
Rhön/Wasserkuppe - 6, 34
Rivals of the Air - 7

S
SK+GA - 28
SS - 14
Scheidhauer, Heinz - 34
Schenck, Sargent - 13
Schmidt, Dr.-Ing. Paul
Sea Eagle - 7
Secondo Campini - 20
Selbstopfermänner - 15
Skorzeny, SS Hauptsturmführer Otto - 3, 4, 6, 9-10, 15-16, 21, 30, 41-42, 45, 48-52, 57, 62, 82
Stab/K 63 - 28
Starbti, Leutnant - 13
Stuka - 7
Suicide Squadron - 33, 61
Surrender, German Unconditional - 7

T
Technical Museum-Delft, Holland - 68
Third Reich - 9
Type 4 Mark 1 Model 20 solid fuel rockets - 17
Tsu-11 - 20

U
Udet, General Ernst - 7-8, 30, 34
Ursinus, Oskar - 34
USS Essex - 9, 11-12, 15

V
Volkswagon - 57
von Greim, General Robert Ritter - 8, 34

W
Walter, Helmuth - 23-24, 26
Wolf Hirth's Grunau Gliding School - 6
Wright-Patterson Air Force Base - 97-98